RUFUS POLLOCK

Dr Rufus Pollock is a researcher, technologist and entrepreneur. He has been a pioneer in the global Open Data movement, advising national governments, international organisations and industry on how to succeed in the digital world. He is the founder of Open Knowledge, a leading NGO which is present in over 35 countries, empowering people and organizations with access to information so they can create insight and drive change. Formerly, he was the Mead Fellow in Economics at Emmanuel College, University of Cambridge. He has been the recipient of a $1m Shuttleworth Fellowship and is currently an Ashoka Fellow and Fellow of the RSA. He holds a PhD in Economics and a double first in Mathematics from the University of Cambridge.

RUFUS POLLOCK

THE OPEN REVOLUTION

A / E / T PRESS

Thank-you for reading. Please share this book and its ideas. We will only realise an Open world when more people are present to its potential. This book is itself openly licensed so you are free to share and reuse it however you wish! The latest digital versions can always be found on https://openrevolution.net/

I'd love to hear what you think of the book. You can share your thoughts via any of the routes listed on https://openrevolution.net/contact

If you got value from the book and have not already purchased a copy, I invite you to make a contribution via https://openrevolution.net/pay-what-feels-right – remuneration rights don't yet exist and your contribution helps us sustain our work.

To my parents

Contents

He who receives an idea from me receives instruction himself without lessening mine; as he who lights his taper at mine, receives light without darkening me.
— Thomas Jefferson to Isaac McPherson, 13 Aug 1813.

Dante: "How can it be that a good when shared, shall make the greater number of possessors richer in it, than if it is possessed by a few?"
Virgil: "Because thou does again fix thy mind merely on things of earth, thou drawest darkness from true light . . . The more people on high who comprehend each other, the more there are to love well, and the more love is there, and like a mirror one giveth back to the other."
— Purgatory XV.

This book is about enlightenment.

Prologue: Monopolies of Attention

In March 2018, when the scandal broke around the political consulting firm Cambridge Analytica and Facebook, the *Guardian* in London quoted a former director of the consultancy:

> Corporations like Google, Facebook, Amazon, all of these large companies, are making tens or hundreds of billions of dollars [from] monetising people's data ... I've been telling companies and governments for years that data is probably your most valuable asset. Individuals should be able to monetise their own data – that's their own human value, not to be exploited.

Other commentators agreed: the problem with these internet giants is their control of our personal data. But this diagnosis is fundamentally mistaken, and, just as in medicine, misdiagnosis matters.

It is not your data that Google and Facebook are exploiting: it is your *attention.* It is your eyes, glued to the screen, that make them all the money; it is because when we want to search for something, or make contact with friends and find out what's going on, *billions* of us turn to these sites. And it is this overwhelming dominance that is so powerful.

We all know that these companies use our personal data to target ads at us, and yes, that is part of the business model, but even if they had no access to data about us they would continue to make huge amounts of money, just as television networks made fortunes before ad-targeting was even invented, simply from the sheer size of their audiences.

It is the *monopoly of your attention* that matters. And so, to diagnose the true problem to which these businesses present, we must ask *how* they have become such incredible monopolies. The answer is that they operate where three different phenomena converge:

1. "Platform" effects
2. Costless digital copying
3. "Intellectual property" rights

It is only when we understand all three of these and their interaction that we have a true diagnosis of the problem – and hence a suitable treatment.

1.1   Platform effects

Twitter, eBay and the others such as Google and Facebook operate as what economists call "platforms", places where different participants connect. This is an ancient phenomenon: the fish-market in the town square is a platform, where sellers and buyers congregate. Amazon does the same, for a wider range of goods and without the smell and noise. Facebook is also a platform, originally designed to connect one user with another to exchange content, though it soon evolved to attract advertisers as well, because they want to connect with the users too. Google is another platform, connecting users with content-providers and advertisers (just as newspapers, for instance, have always done).

All platform businesses have a strong tendency to converge on a single winner. This because the more customers there are, the more suppliers are attracted, and vice versa. For instance, it is strongly in the interests of both buyers and sellers that eBay be as large as possible, so that everyone knows it's the place to come to find what you want. And this mutually reinforcing effect means that rivals are excluded, either deliberately by the company or simply by the logic of platforms working itself out. New entrants cannot compete on equal terms, and so small initial advantages lead to entrenched monopolies. The market converges on a single or a small number of platforms. It worked over centuries for fish-markets and stock-exchanges, and now it works for Google and Facebook as well as Microsoft, Uber and Airbnb.

1.2 *Costless copying*

The owners of fish-markets and stock-exchanges make very good livings. But the owners of the vast online platforms are in a different league because of one of the fundamental characteristics of the digital age: infinite, costless copying. When you start to glimpse the extraordinary ramifications of this simple fact, you begin to understand the modern world.

Once I have a single copy of a piece of digital information – whether it's software, a set of statistics or a symphony – I can make as many copies as I wish, effectively at no cost, at the touch of a button. This is unprecedented. Each new copy costs nothing, since there is no need continually to buy raw materials or new shops from which to sell things. Expansion is free, with infinite economies of scale. So Microsoft, Facebook, Google and the others have been able to scale up their services at an unprecedented rate, and have made unprecedented profits.

1.3 *"Intellectual property" rights*

But costless copying would not be so profitable if it were truly unlimited. If anyone receiving a copy of Microsoft Windows could make as many copies as they wanted and share them then Microsoft would not be able to charge very much. Or if the algorithms that run Google and Facebook were available for anyone to use and modify then other firms could easily compete with them. So the final element that makes these businesses such powerful monopolies is their exclusive right to make the copies. Thanks to "intellectual property" in the form of patents and copyrights, they have exclusive control of the digital information at the heart of their businesses: the software and algorithms that power their products and platforms.

Microsoft Windows is an operating system platform used by much of the world. As an industry standard, it was for a long time effectively a monopoly. But it is only one of the biggest money-spinners of all time because patents and copyrights prevent anyone else from offering its proprietary software for sale. Even though the bits that make up its software and protocols can be copied at no cost, each customer pays tens or hundreds of dollars for the privilege of getting a copy – and this privilege is now almost a requirement for involvement in the digital world. So Microsoft effectively charges each of us a fee to use our computers and for entry to the internet.

It is our framework of "intellectual property" that gives a single company the exclusive right to do this. Yet this monopoly doesn't exist in a state of nature: it is the result of copyrights and patents which we as a society have created. Of course, there is a logic to intellectual property monopolies. Even if subsequent copies are cheap, the initial creation of a new movie, a new app or a medicine can be hugely expensive. Intellectual property is one way to pay for this first instance. But, as we shall see, there are other ways to fund innovation, ways to replace patents and copyrights with remuneration rights, preserving the incentives to innovate but without creating monopolies.

1.4   Old Rules in a New World

And the result of running the information economy by the old rules of intellectual monopoly rights is spiralling inequality. In 2016, the eight richest people in the world had as much money as the bottom 50 per cent of humanity – that's three-and-a-half billion people. And of those eight, six were tech billionaires. This is a political timebomb. It is essential we understand the true causes of this unsustainable concentration of wealth and power: the exclusive ownership of digital information in combination with platform effects and costless copying.

We must see the cost in stunted growth and lost opportunities. By nature, monopolists fear any competition that threatens their position and are driven to neutralize potential rivals either by destroying them or by devouring them. Why, other than to protect its monopoly position, would Facebook pay $22 billion for WhatsApp in 2014 (when WhatsApp's sales were just $10 million)? Although the price paid by Facebook is publicly known, the cost in lost innovation and stunted competition is incalculable. It is the consumer, future innovators and society that lose out.

We need new rules for this new digital world. Taking the old rules of the physical economy and applying them in this new digital one makes no sense. Old property worked, but transplanted into this new world as intellectual property it does not. In this new world, intellectual property is intellectual monopoly. Monopolies that are unjustified and unjust, dangerous both to our economies and our societies. We need new rules suited to our new information economy; rules that provide ways to reward innovators and creators whilst preserving fairness and freedom, and which give everyone a stake in our digital future.

Most simply, we need an Open world. A world where all digital information is open, free for everyone to use, build on and share; *and* where innovators and creators are recognized and rewarded.

Here's how.

2

An Open World

Today, in a digital age, who owns information owns the future. In this digital world, we face a fundamental choice between Open and Closed. In an Open world information is shared by all – freely available to everyone. In a Closed world information is exclusively "owned" and controlled.

Today, we live in a Closed world. A world of extraordinary and growing concentrations in power and wealth. A world where innovation is held back and distorted by the dead hand of monopoly; where essential medicines are affordable only to the rich; where freedom is threatened by manipulation, exclusion and exploitation; and each click you make, every step you take, they'll be watching you.

By contrast, in an Open world all of us would be enriched by the freedom to use, enjoy and build on everything from statistics and research to newspaper stories and books, from software and films to music and medical formulae. In an Open world we would pay innovators and creators more and more fairly, using market-driven remuneration rights in place of intellectual property monopoly rights.

As they have improved, digital technologies have taken on ever more of the tasks that humans used to do, from manufacturing cars to scheduling appointments. And in the next few decades "AI" (artificial intelligence) may well be not only driving our cars for us but drafting legal contracts and performing surgery. On the face of it, we have much to gain if machines can spare us tedious or routine tasks, and perform them with greater accuracy. In future, there is the prospect of our each having more time to devote to things that matter to us individually, whether it's bringing up our children, learning languages or deep-sea diving.

The danger, though, is that robots run on information – software, data algorithms – and at present the "ownership" of this sort of information is very unequal. And because it is protected by our Closed system of intellectual property rights, it is becoming ever more so thanks to costless copying and platform effects. With the overwhelming and ever-growing importance of information technology in the modern world, the balance of wealth and power is tipping further and further towards an exclusive club. But by choosing Openness we can make sure the future works for everyone, not just the one percent.

Already, the world's principal industry is the production and management of information. And control and the wealth of those processes is dangerously concentrated, and is becoming more so. The five richest companies on the globe are all infotech-based, and they themselves exhibit some of the most unequal ownership structures in the world, with tiny groups of founders and investors owning a great proportion of their equity.

As technology accelerates, new *kinds* of applications and experiences are being born which are likely to have a significant place in our everyday lives, as well as in our economies. Virtual reality, for instance, can now replicate many of our sensations and impressions of the world, and has huge scope in future for recreation, as it has already for various forms of training. It would compromise our freedom if virtual reality were to become the same kind of near-monopoly as, for instance, Facebook. Likewise, the so-called internet of things is quickly growing. Already many appliances such as baby monitors, lighting systems and central heating are connected to the internet, but this is only the start. Over the next few years, as billions more devices are connected, we may see machine-to-machine data outstripping human usage to become the principal traffic on the internet. It would be deeply worrying to have control of this fall to a single corporate monolith.

At its most extreme, the current situation threatens the norms of a free society. Free enterprise and free markets are disintegrating in the face of international monopolies, free choice means little when there is only one to choose between, and even our political freedom and freedom of thought are threatened by powers that have the capacity to shape how we think and act. We should all be concerned by this, and the evidence from recent scandals such as Cambridge Analytica show that these concerns are increasingly shared.

Yet if we were to open up to everyone all the information that is being produced – the software that now runs the world, all the riches and the Closed materials, the world's literature and art and algorithms – then we could democratise the infotech revolution. Remember the plan that Google once had of putting every book in the world online? Even Google couldn't do it because it fell foul of copyright. But the Open model would do this not only for all the books, but for all the music, the news, the astronomy and oceanography, market prices, poetry, drug formulae, classical scholarship – all the knowledge and riches of the world that can be digitized. The value generated by our advances would be shared by all humanity, rather than concentrated in the hands of the few. Openness would solve the problem of these monopolies of information power, promoting competition, providing transparency and increasing the possibilities and incentives for innovation. This new approach would make all patented or copyright materials freely available – whilst also paying their creators more and more equitably.

The opportunity and the danger are both great. Choosing optimism and Openness is one of the most important policy opportunities of the 21st century. It is a chance to transform our societies, to create a future beyond the politics of capitalism and socialism, combining the enterprise of the former with the latter's ideal of fairness: a genuine chance to build a better world for everybody. And the entire, unprecedented opportunity is based on one unique characteristic of our extraordinary new digital technology: cost-free copying.

Physical things have an unfortunate limitation: they can be used for only one thing at a time. A bicycle is a bicycle, and if I am riding it to work, you cannot be riding it to the shops at the same time. Physical things, as economists say, are "rival" in use. This fact is so obvious that we barely notice it, but it is of profound importance. It means the world of physical goods is one of scarcity: all too often there is not enough to go round.

Most societies in the world today have systems of private property based upon this physical fact of single-use. We make the *social* control of physical things exclusive because that aligns with the *fact* of exclusive use. If you own a house you decide who lives in it, and the law is built upon the realities of the world's limited and rival physical resources.

On the whole, this has worked well up to now – usually much better than other systems that have been tried. Because our way of thinking has physical property at its heart, we have sought to include information in the same category, where it doesn't belong, under the banner of "intellectual property". In truth, information is fundamentally different. Its unusual and essential characteristic is its boundlessness, its *non*-rivalry, its capacity to replicate. When you share a joke with friends around a dinner-table they each have their own "copy". As such, information is *not* like and should *not* be treated like tangible property.

Digital technology takes this property of information to another level. Once digitized – whether it is a photograph, an app or a symphony – information can be copied as often as we want and shared with anyone *at practically no cost*. Unlike physical things information can be reproduced miraculously to meet demand, making it different from the entire traditional basis of our economy. In this changed world, we need changed rules. Exclusive property rights made sense for physical property because of their scarce and rival nature: with one user, one owner. But digital information is different and its abundant and *non*-rival nature means it needn't be exclusive, it can be Open.

Welcome to the Open Revolution.

3
Defining Information and Openness

On the edge of the Gobi desert in north-west China is the town of Dunhuang. For hundreds of years it was a major stopping place for travellers on the silk road from Europe to China. Carved into a cliff outside the city is a hidden cave, part of ancient holy site called the Caves of a Thousand Buddhas. The cave was sealed up around 1000 AD when Dunhuang was threatened by the Hsi-Hsia kingdom to the north. Forgotten, it lay undisturbed for the best part of a millennium. Then one day in 1900, a young monk exploring the cliffs accidentally discovered the sealed entrance.

Inside was a treasure trove: more than forty thousand silk and paper scrolls and manuscripts, all perfectly preserved over the centuries by the dry desert air. One of the most precious of these is a paper scroll nearly five metres long, made of seven strips of yellowing paper. On the scroll is a copy of the Diamond Sutra, one of the most important texts of the Buddhist faith. Having been obtained in 1907 by the explorer Sir Marc Aurel Stein on his expedition across the Gobi desert, it lives today in the British Library, and can be viewed online.

The scroll is precious not because of its content but because of its form. Rather than written by hand, the text is printed, using the wood-block printing technique which the Chinese invented a thousand years before Gutenberg. Remarkably, the scroll even gives the date of the printing: 10 May 868. This makes the scroll the oldest printed text we have, and a unique testament to our distinctive human desire to record, preserve and share information.

There is one further important feature of this scroll, found in the dedication at the end: the statement that it is "for universal free distribution". That is, the scroll's text can and should be freely copied and shared. Here then, more than a thousand years ago, on the earliest printed text known to man, we have plainly stated the basic idea of free and open sharing of information. The idea of openness, then, was present as far back as the printed records will take us.

It is also likely that the urge to keep information closed – either secret or otherwise restricted – is equally old, especially when the information has commercial value. "Knowledge is power", the old saying goes, and some of our oldest texts, from Homer's Odyssey to the Hebrew Old Testament, provide ample evidence of the power of keeping information closed – after all, the Trojan horse would have been of little use to the Greeks if the Trojans had discerned its purpose.

But when we talk about "information", do we include every thought in our heads and every word we say? Or do we mean something more restricted: only words and thoughts and ideas recorded in a permanent form?

And what is openness? Is it just the opposite of secret? Must open information be cost-free to the user? What about authorship and credit – can works that are "open" nevertheless require that the creators be acknowledged? And finally what about intellectual property such as copyright and patents? How do these relate to open and closed information?

3.1   What is Information?

When we speak generally about "information", we mean knowledge, news, instructions, factual details, formulae and so on. We would not include, for instance, a tune or a poem. For the purposes of this book, though, "information" has a wider meaning. It is taken to include everything recorded in a digital form or in an enduring form that *could* be digitized (such as books in a library).

In short, everything that is or could be written in any language, including equations, musical notes, Morse code or machine code. So as well as the wiring diagram of an airliner, and databases ranging from the human genome to the location of stars in the sky, it includes everything that can be copyrighted – imaginative works such as music, images and stories – and every invention that can be patented.

This book is about making as much as possible of that information available to as many people as we can, since wealth, information and the opportunity to create them are now profoundly entwined. First, though, an important distinction needs to be made between information that is private by nature and that which is non-private. "Revenge porn", to take an extreme example, can be posted online but the material remains intrinsically private. The same is true of, for instance, personal emails and our holiday photos. And privacy extends beyond information we have created ourselves: it includes things such as our health records, our bank statements, and what we bought at the supermarket.

Nor is it only individuals who have information that is legitimately private: governments and corporations have such information too. A company's internal planning and management would not usually be legitimately or legally available to outsiders, and the same is true, though perhaps more restrictively, of government documents. Sometimes a spy or discontented employee steals private information to sell or publish, but that does not mean that it is legitimately public.

On the other hand, all published books, all Hollywood films, all released recordings are non-private, and they are available to everyone for a price. All drug formulae, all research and all inventions are available to anyone for a price – and are therefore non-private information.

How we manage our private information is an important philosophical, technical, political and legal topic, but it is not the focus of this book. This book is concerned with non-private information, information that could be legally or legitimately sold or transferred to *any* third party. So in this book, "information" means non-private information – which encompasses almost all of our commercially and culturally important information, from movies to medicines and software to statistics.

Today, much of this information is in practice tightly controlled, even though it could be legally and legitimately shared with all. It is restricted by copyright and patent law, which limits use and hinders innovation by artificially raising prices or denying access altogether. It is the contention of this book that all non-private information *can* and *should* be Open information, with innovators and creators paid by mechanisms that are compatible with Openness, such as remuneration rights, rather than by the system of intellectual property monopoly rights we have today.

Consider an academic publisher such as Elsevier, the custodian of thousands of new pages of information every year, most of it generated in publicly-funded institutions, which it keeps rigorously closed, behind paywalls thousands of pounds high.[1] Cleverly, Elsevier has inserted itself as an intermediary – a platform – between academic authors and academic readers, controlling many journals which are mini-monopolies in their fields. Increasingly, publishers like Elsevier are exploiting the very academic community they should serve, using monopoly power to hike prices year after year. Meanwhile, they depend for their content and much of the editorial work on the same scholars, who offer their publicly-funded labour (and their copyrights) for free. And since academics have little choice, because they are obliged to publish in "reputable journals", they are held to ransom as surely as the libraries that are obliged to subscribe to the journals.

These monopoly practices are bringing academic publishing into disrepute, and the dam seems likely to break simply because Open publishing provides a flexible, modern, non-monopoly alternative, with the advantage that articles can be readily updated. Meanwhile, consider not only the total of £20 billion global rev-

---

[1] The annual library subscription to a single online journal (usually quarterly) can run to tens of thousand pounds. When Stephen Buranyi wrote an authoritative survey of the field in the *Guardian* (27 June 2017), Elsivier told him that they published 420,000 articles a year, and that "14 million scientists entrust Elsevier to publish their results, and 800,000 scientists donate their time to help them with editing and peer-review".

enues generated by science publishing, but the opportunity costs that result from this mass of information *not* being Openly available for all to build upon. Think of the papers not written and the breakthroughs missed or delayed because scientists are forced to publish their work in journals that lock it away.

3.2 *What is Openness? Freedom to use, build on and share*

What, then, does "Open" mean? Well, Open information has to be more than merely available. It is information that can be *universally and freely used, built upon and shared.*

All three of these stipulations are essential. For information to be regarded as Open, it must first be accessible to all of us to use without payment. Secondly, we must be free, both technically and legally, to build upon it without restriction for our own purposes. And finally, we must be able to share the information, and anything we have built upon it, with everyone else.[2]

Building upon information to make something new is fundamental to our entire culture. Almost no one ever makes anything truly from scratch. Every writer uses techniques learnt from other writers (not to mention his components, the words bequeathed to us by countless generations). All painters learn from other painters – whether imitating or reacting against them. Learning how to do something means learning to adapt existing ideas in new ways. Practically everything we use in everyday life has been designed and made by someone else, and they too were collaborating. Entirely original and independent creations are astonishingly rare. As Isaac Newton stated, "If I have seen further it is by standing on the shoulders of Giants."

Technology is the same, but with the dependency even more apparent. Smartphones, for instance, combine thousands, even hundreds of thousands of ideas and innovations, big, small and microscopic, accumulated over decades and even centuries. As

---

[2]Use, reuse and redistribution are the three core features of openness as set out in the Open Definition https://opendefinition.org/

each is incorporated, the technology advances, allowing them to connect to a cellular network and transmit data thanks to cellular network towers dotted around the landscape and connected by fibre optic cables. The ideas combined in all this have been contributed by people with all sorts of different skills and knowledge.

How many innovations are involved in smartphone technology is impossible to say (how far back do you go?), but we could add up the number of patents involved. To make the implementation of patents practical when so many are used at once, they are aggregated into what are called patent pools, which enable a manufacturer to pay a single licence fee which is then shared out. Naturally, patent-holders are keen to have their patents in the pool, while those with patents already in it tend to oppose new entrants, because more patents may mean a smaller fee for each individually, or make a project too expensive to proceed with. So after this jostling, how many patents are there in the patent pool for 3G? More than 7,500. That is, 3G combines more than seven and a half thousand technological innovations that still have active patents, and which are therefore less than 20 years old. If we were to include older patents and inventions, the numeric keypad, for instance, or the production of the many kinds of plastic, the number of innovations used by 3G technology would be incalculable.

To be Open, information such as that covered by all these patents must be *freely* and *universally* available to use, build on and share. The two qualities of freedom and universality go together, each reinforcing and expanding the other. The freedoms to use, build upon and share must be available to all, irrespective of borders, wealth or purpose. For example, information is not Open if it is available only to those in the United States, or if it may not be used to make a profit – or even used for military purposes. Distasteful though it may sometimes be, universality is especially important to the idea of Openness. An inventor may not want his speech-recognition software to be used to power drones that bomb people. However, rather as if Apple issued an edict that its computers were not to be used for trolling on the internet or posting terrorist videos, this would be impractical and unpoliceable. The power of Openness, like that of freedom of speech, lies in its being available, whatever people wish to do with it. To allow a myriad of restrictions would be to make the system unwieldy and the accumulation of specific conditions would be highly detrimental to creativity.

3.3 Attribution, Integrity and Share-Alike

While Open information must be available for everyone to use, build upon and share, three important provisos can apply: attribution, integrity and an insistence that what is shared must be shared alike.

A creator may insist upon attribution. This simply means that credit must be given in an appropriate way to the author or authors of a work, be it a song or a piece of software. We are familiar with this: novelists, composers and photographers are all credited, and patents list their inventors. Nor is authorship the only kind of credit. Film credits tell us not only the author of the original book, but the names of the director, the actors, and the many others who have contributed (sometimes down to the intern who made the tea). Newspapers attribute the statistics they use, not only for legal reasons but because readers want to know the authority being cited. Listing creators and giving sources, in other words, is a way of accrediting material. Most elaborately, in published papers academics go to great lengths to cite and credit previous researchers and sources, and usually include a bibliography to help others to trace and check them.

Attribution serves several purposes. It offers verification and validation – where did this information come from? where can I see it in its original context? – but it is also a kind of moral recognition: this was made by X, or builds on the work of Y. Credits of this sort, and the reputation that flows from them, are psychologically important and have a practical importance, because jobs and resources are frequently assigned on the basis of achievements and reputation. This factor is even more important when the creator earns little or nothing directly from his work, as in the case of a mathematical breakthrough or a scholarly paper newly attributing a sketch to Constable for the first time. This is particularly true in the case of Open materials, freely distributed, as increasingly they are on the internet. The requirement for attribution usually places little or no burden on those who use, reuse or redistribute information.

The second stipulation that Openness permits is a requirement to respect integrity. "Integrity" is a current legal term in the regulation of information, and refers to the control that creators may exert over the way their work is used or altered (whether it has been freely obtained or paid for). Integrity arguments were deployed in 2006, for example, in an attempt to prevent female actors taking the leading parts in an Italian production of Samuel Beckett's *Waiting for Godot*. But since this right can be used to block new uses of a work, it is at odds with the freedom and universality that are at the heart of the philosophy of Openness, and they will be more narrowly interpreted in the Open world. If information is to count as Open, the integrity requirement must not grant the original creator a veto power over changes by reusers. Others must be free to use the work for their own purposes. There may, however, legitimately be a requirement both to declare and to explain the relation to and differences from the parent.

The third stipulation that Openness permits is for share-alike, requiring that those who reuse work that has been freely shared must in turn share their own work Openly in the same way – and with a share-alike requirement in turn. In this way Openness cascades down the generations of creativity.

Share-alike is most significant in areas where reuse is common. The concept of "share-alike" originated in the 1980s with the work of Richard Stallman in the building of software, where reuse is ubiquitous. His concern was that if he shared his work freely and Openly, others might take it and copyright it rather than sharing in their turn. Share-alike requirements solve this problem, and the beauty of the system is that it imposes no burden on those who are sharing. But it has a ratchet effect that can bring more and more material into the Open realm. Everyone who uses this material must adopt the share-alike system, and so on unto the third and fourth generations. And share-alike is already required by many major Open information projects such as Wikipedia, OpenStreetMap, GNU/Linux and Android.

None of this, however, means that Open publication is sheer altruism, giving away one's work for nothing. There are mechanisms by which Open publication can be rewarded – and in fairer and more socially beneficial ways than it is at present. This too is part of the vision of Openness. But first, where do we stand now?

4
Patents and Copyright as "Intellectual Property"

Of the two major kinds of monopoly rights over information, patents are regarded as the broader, covering more of the idea or approach of an invention, whereas copyright focuses on exact, or close to exact, duplication. Originally concerned with the copying of printed books, the scope of copyright has grown to include almost all material that has a precise linguistic or symbolic form and can therefore be copied. This now includes not only cultural works such as music and films but commercial information such as software. And the distinction from patents has become somewhat blurred as copyright has been extended to cover, say, fictional characters and the design of software interfaces. Whereas patents and copyright were originally distinct, they are now classified as branches of the same tree of information-related monopoly rights, "intellectual property".

Nevertheless, distinctions remain. For example, patents are relatively short: even if extended, they rarely run for more than 20 years. Copyright, by contrast, is now very long: often amounting to a monopoly for seventy years or more beyond the life of the author, so that it is common for the works of authors long dead to remain under the control of descendants or trustees, or of corporations which have bought the rights.

There are, in addition to copyrights and patents, other information rights under the heading of "intellectual property". The most significant are trademarks – which are essentially rights to control branding – but there are also laws about trade secrets and some much newer rights such as those to do with databases.

Yet even with all of these forms of "intellectual property", the rewards for inventiveness are not comprehensive, fair or proportionate. Most obviously, the inventors of many everyday things receive no rewards at all because they are concepts rather than products, means of solving problems or even forms of behaviour which may benefit billions of people but have no financial value. If, for instance, had you been the first to invent the roundabout, so easing congestion all over the world, you would have had nothing to patent. If you had solved a conundrum in mathematics or physics, you would not have monopoly control over the solution, nor immediate financial rewards from it. If you write the words to a song, you have a copyright. If you write the music, you have a copyright. But if you invent the strobe effects that go with them, you have nothing.

Although the history of patents and copyrights is long and tangled, generally involving extensions of both their scope and their duration, they were from the first designed as monopolies, and neither was initially construed as property in the way that the modern term "intellectual property" invites us to do. Yet in the past few decades, advocates for patents and copyright have increasingly sought to free them from any negative association with monopolies, and to replace this with a positive association with the much more palatable idea of private property.

The adoption of the language of "intellectual property rights" to designate legal exclusivities regulating the flow of information is not innocent. This branding has associated these restrictions with traditional rights in tangible property, which are generally respected because of custom and experience (and perhaps especially the experience of the past hundred years, during which private property has so often been catastrophically violated). This deliberate confusion amounts to a rhetorical hijacking.

Information is *not* the same as tangible property, since it is not rival or exclusive in use. Information is not naturally *property* at all. You cannot *possess* Mozart's last symphony or Fermat's last theorem or the rules of chess: once created, they float free. They belong to us all.

The same is intrinsically true of a new tune, a furniture design, or a novel. But not in law, where these are initially subject to the monopolies of copyrights and patents, which offer creators and investors a means of profiting from their efforts and risk-taking, and so give an incentive for further production. These restrictions, however, work by *limiting* people's access artificially, inflating prices, and curtailing the scope for third parties to reuse the information in their own work. We should not bamboozle ourselves by confusing the rival nature of tangible property with the deliberately imposed monopolies that restrict our access to and use of information.

In his novel *The Man Without Qualities*, Robert Musil writes that "fire does not become less when other fires kindle from it". The same is true of information: it does not become less when others Kindle from it. The formula for Ibuprofen and the source code for Linux aren't threatened by over-use. And given the non-exclusive, non-rival nature of information, the natural way to treat it is the Open model – a collective commons to which all have access. In fact, there are times when a good portion of humanity is enjoying the same information at once, such as when we all watch the final of the World Cup or the Olympic 100 metres, and the sharing is itself a vital and enriching part of the experience.

As well as being in accord with the nature of information, free sharing is the only way for society as a whole to realize its full benefits. But if there were no commercial incentive to create such informational goods, we might not develop them in the first place. Once made, a Hollywood blockbuster can be copied across the internet in seconds for almost nothing, but creating the master video file may take years of effort and tens or hundreds of millions of dollars. So there is a tension between allowing the Open sharing of information and the need to pay for that expensive original. If the film had no copyright protection and were free to copy, where would we find the resources to make it in the first place? And who would fund our billions of dollars' of medical research if the cures that eventually emerge were not protected with monopoly rights over them as "intellectual property"? Huge interests are at stake, and so are reputations and livelihoods.

Of course creativity and innovation should be recognized and rewarded, but exclusive rights are not the only way this can be done. Until the mid 18th century, for instance, many writers and composers were rewarded by the patronage of royalty and the aristocracy. Nowadays, many branches of the arts and sciences are fostered by patronage of other kinds, whether it be through commercial sponsorship, university funding, the Arts Council, or, particularly in America, great trusts and foundations. Creators are often rewarded indirectly by the recognition of their achievements, and being sought-after for their celebrity. Einstein didn't have exclusive rights over his ideas. People do not pay to use the theory of relativity or the formula $E=mc2$. He published them Openly for everyone to read, analyse and build upon. Most of his career was paid for by universities, whether publicly or privately funded.

Patronage has the disadvantages that you may back the wrong horse and that it creates dependency upon the rich. Market mechanisms, on the other hand, provide opportunities for everyone, and specifically reward innovations according to take-up. This in itself has disadvantages – particularly, as we have seen, if the creator has a monopoly. But there are better ways to fund the creation of information than by imposing exclusionary monopolies. By banding together – usually through our taxes – we can raise the money to pay for information goods as they are created, much as we raise money to pay for national defence or roads. Moreover, we can do this, if we want, without removing any choice from consumers or freedom from the market. The state can coordinate the raising of money but leave the market and entrepreneurs to decide what information is created and consumed: which movies are made, which lines of medical research are pursued, which software is written.

No one wants to see a government committee deciding which authors to support or what software should be written, but traditional, demand-driven market mechanisms can be used to allocate all or part of the money collected. Rather than the patent and copyright monopolies they have today, innovators and creators can be given "remuneration rights". These would entitle the owners to payment from a remuneration rights fund, according to the value that the information generates – for example, how much impact a specific medicine has on improving health or how many times a song is played.

———————————————

So is an Open Revolution possible? Yes, and it is based on solid experience and statistics. More and more of the world's software is Open. Four out of five smartphones run on an operating system that is Open and free, developed by thousands of organizations and individuals over more than forty years and freely shared. And this has occurred without the sort of systematic public funding that exists, for example, in science. It has also happened without the supposed financial advantages of proprietary software. Almost every aspect of the Open approach has been tried successfully in one area or another.

1. The internet itself, for instance, is an amazing example of an Open platform that can be used by everyone. On it, you can already find huge amounts of Open-source material available not only to use but to build upon.

2. We already use our taxes to pay for some of our information. The BBC, for instance, is paid for by a licence fee by all who watch television, but it is not a monopoly, and is legally obliged to commission a proportion of its production elsewhere. Close to half of all medical R&D in the United States is funded directly by taxpayers.

3. Mechanisms such as collecting societies already distribute revenues from recorded music according to air-plays and downloads. The Spotify and Netflix approach of a fixed fee and unlimited access has much in common with an Open approach.

4. From different countries at different times, we have examples of what happens to, for instance, the medical market in the absence of provision for drug patents.

5
Face to Face with Power

Although the platform monopolies Google, Facebook and Microsoft each consist of a single firm, this need not be so. It is possible for a platform to be neutral, either not owned by anyone or owned by all of those who use it. For its computing and communication needs, the world has converged on a single network and single set of protocols, yet the internet is not owned or controlled by any one firm. As long as you adhere to the technicalities of the internet protocols and certain legal rules excluding anti-social content, you can connect to the internet and to other users. The internet is a platform that mediates between all of its users impartially.

Contrast this with Facebook and you can see how different things could be: Facebook provides media sharing, communication, identification and spam-management services, but its protocols and platform are largely proprietary and controlled by the company, which ultimately determines who uses them and for what. The difference between these two kinds of platform was made starkly clear in spring 2018 by the Cambridge Analytica scandal. Facebook took a large share of the blame for misuse of personal information; no one blamed the internet itself.

There is no reason, though, why Facebook could not have been like the internet, with its protocols being Open and universally accessible. Instead of a proprietary social network controlled by one corporation, we could have had an Open social network, owned and controlled by its users – just like the internet itself.

In an Open social network, anyone – suitably identified – could connect and innovate on the platform. However, as things stand (and although we may not realize it), Facebook is able to exclude anything that might impinge upon or threaten it. You would not, for instance, be able to use it to build your own social network, or to introduce a plugin that blocked Facebook ads.

And Facebook's power extends far beyond its own web pages. On 2 November 2010, the day of the US Congressional elections, Facebook placed on the newsfeed of its 61 million American users an informational message about voting, together with an "I Voted" button, allowing friends to signal to one another. Facebook's intention was innocent and involved no deliberate partisanship. Nevertheless, the results were striking. Analysis reported in 2012 showed that Facebook's move accounted for probably at least an additional 340,000 votes. This was 25% of the entire increase in turnout, making Facebook the biggest single factor affecting increased turnout. Whilst this change may not sound very significant, additional turnout can be crucial. For example, the 2000 election between Al Gore and George W. Bush was ultimately decided in Florida by a margin of 537 votes – less than 0.001% of all voters. In 2016 a shift of a mere 100,000 votes overall from Donald Trump to Hillary Clinton in Pennsylvania, Michigan and Wisconsin would have made Clinton President. Facebook's interventions in 2010 were very small, just a single message and a button. More concerted or targeted efforts such as those by Cambridge Analytica can have a much larger impact.

Facebook repeated its experiment in 2012, but the results have not been published. Facebook is probably wary of sharing its work publicly following the reaction in 2014 to the publication of the results of its "emotional states" experiment, in which it found that adding more negative or more positive items to some users' newsfeeds appeared to affect their emotions. Facebook has also investigated how prominent inclusion of "hard" news might influence voter turnout, but has not published the results. One need not question Facebook's present intentions to find this a cause for concern. The huge potential power of such platforms is now undeniable, and this power *could* be used deliberately for political ends.

And it is not just Facebook. A study published in 2015 in the Proceedings of the National Academy of Sciences showed that Google has the power to shift elections through its ability to shape the search results it delivered for a politician or political party. In a simple experiment with real-world voters, researchers demonstrated that manipulating searches to provide more positive or negative results had a significant impact, especially among the undecided. Furthermore, they showed that this effect would be sufficient to change the results of many elections around the world including recent close elections such as the US presidential contest between Trump and Clinton.

While advanced democracies generally take measures to ensure that media ownership does not become too concentrated – Germany, for example, has explicit limits on the percentage of readers or viewers that any one company may have – platforms such as Facebook or Google have achieved a dominance of users' attention far greater than almost any newspaper or broadcaster in history. This near-monopoly power in social media is greater for being less explicit and obvious, with scant transparency or oversight, and it has the potential to limit our freedom of speech, enquiry and even thought.

Action is clearly needed. But what exactly can and should we do? Simple regulation seems both insufficient and unsustainable – what happens when the next Facebook appears? Moreover, regulation risks entrenching these monopolies further: well-meaning oversight can rapidly turn into rigid rules that form an insurmountable barrier to new competitors whilst little impeding the monopolist (see below for the example of the FCC and AT&T). In addition, traditional regulation entails bureaucratic oversight which may struggle to keep pace with innovation. Fortunately, the Open model provides an alternative way forward. It offers a solution to the problem of monopolies that avoids heavy-handed regulation and fosters sustainable innovation and free competition. By using remuneration rights to pay innovators, we can combine the power and discipline of the market with the openness of the internet.

6
Triumph over Closed Minds: The Internet

The internet is *the* infrastructure of the information age. It is the road and rail of the modern era – an information superhighway. It is also the greatest example of an Open system that we have, created to an Open design to provide freedom and possibilities to all. The freedom it gives has allowed the global community to create new uses unimagined by its original architects. It was on the internet that Google and Amazon were launched; it was on the internet that a billion websites bloomed; it was on the internet that the digital economy took off.

Openness was central to this prolific activity, but it was by no means inevitable – in fact, it is an anomaly. Other telecommunications networks have almost all been Closed, having both a restricted series of connections and specific, limited forms of data. Permission to connect was closely guarded by network owners – think of national telephone monopolies – and the uses to which they were put were predetermined.

In 1992 it cost $5,000 to buy the Blue Book, the manual containing standards for the world's telephone systems (published by the ITU in Geneva). When it came to mobile telephony, the principal standard from the early 1990s to the mid-2000s was GSM, and if you wanted to build your own GSM system – for example, a base station for receiving signals from your own handsets – you would need to understand how GSM worked and have permission to use that information. Neither was possible: information on how a base station worked was strictly controlled (it was not until

2010 that a white-hat hacker managed to get hold of a GSM base station on eBay and reverse engineer the protocol). In any case, permission to use that information, even if it had been available, was restricted by a large number of patents.

As for connectivity, for most of the 20th century telephony was regulated by governments through effective monopolies, which did their utmost to resist encroachment, as in the David and Goliath story of Henry Tuttle and America's gigantic corporation AT&T.

Mr. Henry Tuttle was the proud inventor of a telephone silencer, unpromisingly called the Hush-a-Phone. It was a large plastic cup that you attached to the speaking end of a telephone handset so that no one around you could hear what you were saying. Mr. Tuttle had been in business for years when, in the late 1940s, he received the alarming news that devices like his were to be forbidden by AT&T, on the basis of an obscure provision of its agreement with the government which stated:

> No equipment, apparatus, circuit or device not furnished by the telephone company shall be attached to or connected with the facilities furnished by the telephone company, whether physically, by induction, or otherwise.

Put simply: you could not connect anything even to your own handset without AT&T's permission, presumably including a plastic cup from your picnic basket. AT&T's network was at that time and for long after the only large-scale communication network available, so it was pulling the plug on Mr. Tuttle's perfectly harmless business. In 1950, AT&T took Tuttle to court, or rather to its equivalent in this area: a special hearing of the regulator, the Federal Communications Commission in Washington DC. One might have imagined this was a minor affair concerning an obscure product not in competition with the phone company and bought by a tiny minority of its customers. But AT&T showed up in force: dozens of attorneys plus a bevy of expert witnesses and top-level executives. Tuttle had only himself, his lawyer, and two acoustic professors from Harvard.

For AT&T this case was not about the Hush-a-Phone but about

the principle of connecting to the phone system. And behind that, something much bigger was at stake: who had control. For if Mr. Tuttle were allowed to do what he did, then they might also have to allow all sorts of other innovations to be connected, and if lots of people could do things uncontrolled and unsupervised by AT&T, there was a potential danger. Someday, someone might invent something that would disrupt its business. A company that was sitting on one of the safest, soundest, government-guaranteed monopolies in the world did not want anyone muscling in on any part of its operation, however peripheral, and even if it didn't offer any equivalent.

After a five-year delay, the FCC issued a ruling that the Hush-a-Phone was indeed "deleterious to the telephone system and injures the service rendered by it", so AT&T had the right to forbid it, and devices like it. And there the case might have rested, so dooming the internet even before it was conceived.

Really? The internet and the Hush-a-Phone are utterly different. Yes, but they share two key common features. First, both can be construed as attachments to the phone network. No one was going to buy a Hush-a-Phone silencer without a phone that could make calls. And the internet has to send its data either down wires – metal or optical cables – or as electromagnetic waves through the air or through space. To create the internet, access to a transmission system was essential, and in the late-20th-century America, AT&T's was the only one. For it was AT&T that had run millions of miles of copper wire to reach almost every home and business in the land, and connected each of its local networks into a national network, making enormous investments in high-capacity long-distance lines. It would be decades before any real alternative arose locally in cities, with the systems installed by cable TV companies, and then for long-distance transmission, first from satellite providers and later with fibre-optic lines financed by the internet boom. Even today, more than fifty years on, AT&T and its equivalents in other countries are often still the only providers of "last-mile" connections into homes and businesses. In sum, the internet, or any service like it, could operate in the US only with access to AT&T's network.

So AT&T went to extraordinary lengths to suppress the innocuous Hush-a-Phone. Such devices, they claimed in evidence, posed a threat to the safety and functionality of their network, and vivid pictures were painted of repairmen being injured or electrocuted. And if they thought a plastic cup could be fatal, imagine the conniptions they would have had if they had dreamt of something like the internet. For this would mean using AT&T's wires to send entirely new kinds of signals and messages. Not just pieces of imitation crockery but entire computers were to be attached to AT&Ts lines.

Of course the FCC couldn't know that its anti-competitive ruling of 1955 might have the effect of muffling not only the humble Hush-a-Phone but also the greatest technical breakthrough of the following half-century. That is one of the great ironies and challenges of innovation policy: we don't know what we don't know.[1] The future has no lobbyists or lawyers. We inevitably make decisions based on what we can anticipate or imagine, but innovations of the most important and exciting kinds are often about precisely what we cannot anticipate. This is the reason, worth reiterating, why Openness is so crucial: an Open system or platform allows anyone to build on top of it, and so allows the maximum variety of innovation.

Fortunately, though, Henry Tuttle was not, so to speak, to be silenced. He was indignant and determined to press on, despite substantial costs already incurred over several years. He went to the Court of Appeals, and in 1956 a panel of federal judges headed by Judge Bazelon unanimously overturned the FCC's decision. Furthermore, at the end of their judgment, in its penultimate sentence, there was a crucial phrase. An AT&T user had the right "reasonably to use his telephone in ways which are privately beneficial without being publicly detrimental." That one phrase was to give the internet the opening it needed. It punctured the Closed system that AT&T had been defending, and AT&T was right to be afraid, for a few years later the competition came

---

[1] When asked what would be the use of the newly-discovered electricity, Michael Faraday is said to have replied "What's the use of a new-born baby?"

pouring through that breach and its once all-powerful empire was to be fundamentally and fatally undermined.

The two expert witnesses in the FCC's Hush-a-Phone hearing in 1950 were the Harvard acoustics professors J. C. R. Licklider and Leo Beranek. Both were to play central roles in the establishment of the internet and its distinctive Open philosophy. In the late 1950s, Licklider became fascinated by computers and the problem of how to improve interaction between them and humans. In 1962, he was appointed head of funding in the area of computing at the Pentagon's Advanced Research Projects Agency (ARPA). Suddenly, he controlled a bigger budget for computer science research than the combined budgets of all other such efforts in America, and he used it to fund some of the most imaginative blue-sky research of the time. Central to his vision was the idea that if computers were truly to enhance human thinking, ways had to be found to communicate *with* them and *between* them. The earliest glimmerings of the possibilities of networked computers had been identified.

Licklider stepped down after two years at ARPA, but his ideas gathered momentum thanks not only to his successor, Bob Taylor, but to Paul Baran's packet-switching ideas at Rand Corporation, and to many others. In August 1968, the tender went out to build the first prototype implementation – to be called the Arpanet – which a few years later became the seedling of the internet we know today. The contract to build it went to a small consulting firm with a reputation for brilliance and informality, Bolt, Beranek & Newman (BBN), founded by Licklider's old Harvard colleague, Leo Beranek.

This isn't a history of the long road from that prototype to the internet we have today, with its billions of daily users and traffic measured in petabytes. The important thing here is the philosophy that the internet enshrined, which is so different from any other communications network existing then or since: Openness of access and information.

Here's one extraordinary example: while Bolt, Beranek and Newman were building the initial four-host network in 1968, questions arose as to what would actually be sent over it. Email and web pages had never been thought of. BBN were responsible for creating reliable physical links across which data could be sent, but what data would this be?

What happened? An interested group of graduate students at the universities involved spontaneously formed, calling themselves the Network Working Group. They contacted BBN and were given informal approval. They started publishing ideas and specifications under the rubric "Request for Comments", which emphasized their informality. Between them, they invented what became the protocols of the internet, publishing them early, often and Openly. To understand how remarkable this was, you have to remember that ARPA was a sub-agency of the Department of Defence, where contracts conventionally went to big firms with staid bureaucracies (the major bidder against BBN had been the huge defence contractor Raytheon; the computer corporations IBM and CDC had refused to bid, because they thought the project must fail). In addition all telecommunications in the United States were run by AT&T, the most hidebound and hierarchical of corporations. At AT&T, graduate students would not have come anywhere near this project, let alone been left to design the core specifications in a completely Open forum. Every single major specification of the Arpanet – and hence the internet – was hammered out informally without any official committees. You could get every single specification for free as Open software from the start. (Readers over a certain age may remember how astonishing it was that to begin emailing one had to do no more than establish an account and press "send": no fee, no licence, no official carrier. Magic.)

Contrast this with the traditional telephone industry, where standards were arrived at in committees working for years, and access to them was limited to the priesthood. Not only were the specifications Closed and closely controlled by the International Telecommunications Union (ITU), but in 1992, when the Blue Book could perfectly well have been put on the internet for free, it still cost $5,000 and was locked up in software so old that the ITU itself could not read it properly.

By contrast, thanks to the influence of Licklider, Baran and others, the internet had an Open architecture. There was no central control, the network was distributed, and anyone could connect anything to it, so long as they followed the protocols. This was so alien to AT&T as to blind them to its import. In 1972, the company was offered the chance to take over the Arpanet, as the fledgling was still known. Senior managers and experts there considered the matter for months, and then politely declined, citing its incompatibility with their network.

AT&T weren't the only ones with closed minds. With the spread of personal computers in the early 1980s, several national telecoms companies created their own mini-information networks, such as France Telecom's Minitel and British Telecom's Ceefax. Several were more sophisticated than the internet of the time: a decade ahead of the World Wide Web, they were carrying into homes real-time information such as weather forecasts and train times. But they differed in one crucial aspect from the internet: they were not Open. The owners alone determined what information found its way onto these services.

Fortunately, the internet beat off the threat from these Closed systems. Thanks to careful nurturing and a strong base in academia, it was resilient, and by the mid 1980s it was poised to take over the world.

How was this possible? How could the internet be so different? Much of the credit must go to the fact of government funding. The beginnings of the internet were almost entirely paid for by government research funds in the US and to a much smaller extent in the UK (where work at the National Physical Laboratory under Donald Davies was crucial to the development of packet-switching). Even more important was the form of the government funding. Today, ARPA is a legend of what is possible for a public agency. It was staffed by outsiders and free to make bold bets with a minimum of bureaucracy. Funding from ARPA helped to create not only the internet but other aspects of digital life that we now take for granted, from user interfaces to the mouse. This was money with a vision, and a vision that something unprecedented was possible.

In addition to this essential government support at an early stage, the internet had the good fortune to mature in a relative power vacuum. At first, commercial operators did not understand what was happening and how it would affect their businesses. During the 1960s and 1970s, with the support of the White House, the FCC took an increasingly tough line with AT&T, insisting on greater Openness and competition. As a result, the internet reached early maturity before any one player could try to dominate or control it. If AT&T had retained its monopoly on the digital networks of the US, we would probably have a system of sorts allowing digital communication, but it would have been less like the internet than like France's Minitel service, very limited in scope and perhaps very expensive. Sure enough, when the internet took off in the early 1990s, big players ranging from AOL to Microsoft to AT&T tried to seize control, but they were too late; it was too big for any one corporation – or even government – to own.

———————————

The internet and the web have been the greatest innovation platforms of all time in terms of quantity, quality and velocity of what has been built on and around them – largely because no monopoly controls them. The monopoly issue is a live one, though, because without active efforts to promote Openness, our digital world keeps tending towards proprietary monopolies. Facebook, for instance, has been creating a proprietary layer on top of the internet. More and more people do not log on to the internet, they log on to Facebook. To most of us, including myself, this seems at first to be fairly innocuous: Facebook provides a great service and all my friends are on it. Even as I use it, we are scarcely aware that it dominates our internet usage more and more, whether I am messaging friends, organizing events or posting pictures and thoughts. Subtly and gradually, however, Facebook is becoming where we live online. Some 80% of social traffic now goes through this single company.

Facebook is a monopoly, and its CEO and investors are anxious to keep it that way, so the company has been busy making sure that the innovation that happens around it works not to threaten but to reinforce it. Of course it is difficult to see the negative effects, the companies and innovations that never made it, or which have been absorbed into Facebook and neutralized, diminishing innovation as they disappear.

Imagine you want to start your own innovative social network today. To get started, you almost certainly need to co-operate with Facebook in some way, so that your users can exchange content with their friends and the world whilst on Facebook, rather than on an entirely separate network of yours. But does Facebook have an incentive to make this easy or will it want to hamper your efforts, subtly or otherwise? Alas, we know the answer. This is the great irony: the Openness of the internet made Facebook possible, but that Openness is now a threat and Facebook is gradually rendering the internet Closed.

As an example of a better way, let's have a look at the operation of music streaming and how it might be operated to the benefit of a far wider public – an entire nation at a time.

7
Music to our Ears

The music-streaming service Spotify, founded in Sweden in 2006, enables users to listen to songs over the internet, one after the other, without downloading them. By 2017 it had more than 150 million customers. When describing Spotify it is usual to call it a "streaming" service, but streaming is actually a sideshow, even a gimmick. How, though, does streaming work? Strictly, it means that the information that constitutes the songs is sent to the user in a continuous "stream" rather than having to be downloaded before it is played. The importance of this distinction is that users never have the whole song; they have only the part they need to play at this second. In many ways, it is very similar to radio. Your radio plays only an instantaneous part of the broadcast as it receives it. You can't go back and listen to earlier parts, or receive the whole broadcast and play it when you like. It's obvious, of course, why radio works this way: radio waves are being broadcast to everyone, and radios originally did not have built-in means of recording and then playing back.

Spotify, though, is streaming over the internet where no such limitations apply. It is not broadcasting: music is sent to each individual user. So why not allow each user to download the music, or at least to pause or skip tracks at will (which would be trivial to implement)? There are two answers. Technically, streaming has the advantage that users do not need to wait for the whole track to download before beginning to listen – they can start listening the moment the first chunk of the track arrives.

But this is a very limited advantage. It would be easy enough to queue songs for download in the background instead, so that each was ready as the previous one finished. And there is no technical problem that prevents users from pausing, skipping or keeping copies of tracks to listen to later. The reason for these restrictions is the law of copyright.

Once digital music became available in the form of CDs and the internet had decent bandwidth, the natural thing was to put the music on the internet and let people listen to what they wanted. They could listen for free, because by using peer-to-peer distribution users could simply download it from one another. And this is what people immediately started doing, most famously using the free platform Napster. But in 2001 a court decision in the US declared Napster illegal, on the basis that although it was not hosting any content itself, it was enabling massive copyright infringement by its users, who were downloading and listening to music without payment or permission.

The next generation of companies, such as Spotify and Last.fm, learnt their lesson. Whilst it would have been easier for them to make systems that simply let users download whatever they wanted, they went out of their way to limit this and to make their online services like radio. Why? Because radio has special legal status with respect to copyright, thanks to decades of negotiation and law-making that has led to an accommodation between broadcasters and copyright holders such as recording companies.

Radio stations have blanket licences from "collecting societies", which permit them to broadcast music without obtaining a licence for each piece they play. The collecting societies then divide that pot of money amongst the copyright holders, roughly in proportion to airtime.[1]

---

[1] The situation varies from country to country. In the US, broadcasters pay the composers of the songs but don't have to pay performers. The two copyrights have been distinct since the second was created in 1972, and the exemption is largely due to the power of the broadcasting lobby at the time. Its argument was that broadcasters provide a valuable service for performers by promoting their recordings on the airwaves. The recording companies clearly believe so, because they have spent large

When streaming began, therefore, the companies reasoned that if they could offer consumers something sufficiently similar to radio, they too could avoid negotiating individual licensing deals and instead use a blanket licence from the collecting society. Without that, streaming would have been nearly impossible, and so Spotify, Last.fm and the others deliberately compromised their products by, for example, limiting the number of times a user could skip tracks in a 24-hour period.

Then, once they had started building a user-base and taking money from investors, they negotiated with individual music labels and artists for licences, so that they could start charging users for options such as choosing their own tracks or downloading music to listen to at any time (though Spotify does not have rights to several major artists including Taylor Swift because the copyright-holders think its royalties are too low).[2]

Even the premium version has limitations, however. For example, Spotify is more like a rental service than a retailer: if you stop subscribing to its pay service, you lose access to all the music you have downloaded, which is a considerable incentive to go on paying. This restriction arises both from Spotify's agreements with the labels and from its own interests. You cannot simply download all the tracks you want and then cancel your subscription. As one user notes, Spotify's premium service is addictive. Once you are hooked, it is hard to leave: "I lasted two months and wound up going back to premium: $9.99 is not that much for a crack addiction."

With no fee per track and no limitation on use, this all-you-can-eat buffet is a prototype for how one aspect of the Open world would operate. Money would of course have to be collected somehow to fund it, but instead of ten dollars a month to Spotify, this could be a special fee incorporated in your taxes or added to your internet or mobile bill. This money would then be distributed according to usage, through remuneration rights fees.

---

amounts of money, sometimes illegally, to get prominent stations to play their artists' records.

[2] The Beatles were the most famous hold-out from streaming audio platforms but since 2016 most of their catalogue is now available.

Suppose the Netherlands had made the transition to Openness. Every recording ever made anywhere would then be Openly available within the Netherlands, but with an electronic wall to prevent people in other countries accessing them (at least until those countries also became Open). Any Dutch citizen would be at liberty to listen to, share or remix any recording or composition. To pay for this, the government might, for instance, add a small charge to the data plan of everyone's phone. Unlike Spotify's fee, this would be very low.

How low? Well, let's examine how much it would take to pay creators as much or more than they receive today. Currently, the total revenue for music in the Netherlands is around €150m a year. Of this, probably less than 60% goes to creators, but let's err on the generous side and suppose that all of it does. With some 15 million adults in the Netherlands, a fixed fee per adult to pay for all current use of recorded music would be €10 per person *per year*. At 85¢ a month, this is less than a tenth of Spotify's €10 a month. If the levy were payable only by people with internet subscriptions, the charge would be around €1.75 a month per connection.

Whatever the method to raise the funds for the music industry, there could be several mechanisms for allocating them, which could themselves be combined. Given the importance of individual taste in artistic judgments, they would be weighted differently. Here is a suggested allocation:

- Remuneration rights – say 80% of the funding for music – would be issued for both compositions and recordings, and would entitle each holder to a share of the remuneration rights fund. This would be allocated in approximate proportion to usage of works. A legal and administrative framework for this is already in general use in the industry. For example, composers provide automatic fixed fee licences to recording artists, and collecting societies administer collective licensing for performances to commercial users such as shops, bars and nightclubs. In the Open world, the overwhelming bulk of funding would be distributed this way. One possible change, which has already been pioneered by some collecting societies, would be to make distribution progressive, reducing the proportion paid to the very biggest stars so as to pay more to those earning less, in order to support the up-and-coming and experimental.

- Traditional expert-selected grant funding – say 10% – would be allocated up-front to particular artists or organizations to create new pieces and recordings. This would be similar to the work of existing public arts programs around the world, though it would cover information-production but not live performance.

- User-choice (the "Kickstarter" or "X-Factor" model) – say 10% – would allow some active consumer-choice in the allocation of funding to particular artists, projects or even general policies (supporting blues artists, for instance). Artists would propose projects, such as an album or new song, with a budget. Citizens would each be allocated "voting dollars" with which they could support such projects (with unused dollars being allocated proportionally). This would give the public some control over up-front funding, and has similarities to crowdfunding schemes such as Kickstarter or audience-voting on shows such as X-Factor.

So, you could have an Open music system in the Netherlands and pay artists *more* than they receive now, for less than the cost of a bus ride each month, or a even bottle of water. Perhaps the levy could be put on water in plastic bottles!

Just how great are the benefits of increased access and use? It's impossible to say precisely. How much additional usage would there be; and what is the value of a child having the chance to hear Beethoven's Ninth or a grandmother dancing to the hits of her youth? Yet there are ways to calculate a rough monetary value, and the best and most recent estimate comes from a team led by Professor Bernt Hugenholtz at the Institute for Information Law at Amsterdam University, which carried out a study from 2012 to 2015. Their results indicated that a move to an Open music model using an alternative compensation system would create extra value of the order of €600 million a year for Dutch society

– more than four times the entire annual revenue of the nation's recording industry.[3]

Even this almost certainly underestimates the benefits, because it does not include other gains from reducing or eliminating costs in the current inefficient system. For example, there are all the legal costs of traditional licensing, the enforcement costs for rights-holders in suing infringers, trying to prevent file-sharing, and so forth. Exclusion is key to private monopolies, and the wasted opportunity turns into a waste of money. Users who are excluded will often try to get round the paywall, for instance by asking friends who subscribe to the service to stream it to them in turn, or to download and share it. Since this would reduce its pool of potential customers, Spotify does various things to restrict the service – even, one could say, to cripple it. In particular, it uses Digital Rights Management (DRM), which encrypts all the music it sends to you so that only you can play it. Users cannot play the music through whatever application they like, and are able to copy it only to another device with a Spotify app. Spotify then has to spend time and money defending the DRM against people who want to hack or disable it, and suing anyone who does.[4] All in all, it spends a lot of time and money implementing and maintaining a system the purpose of which is to make its service *less* useful.

Such elaborate digital obstructions are not unique to music. They take many forms. They range from restrictions on which BBC programmes are freely accessible (having, of course, been paid for by the British public) and when, and where, to digital

---

[3] *Going Means Trouble and Staying Makes it Double: The Value of Licensing Recorded Music Online* by Christian Handke, Bodo Balazs and Joan-Josep Vallbé in *Journal of Cultural Economics* 22 May 2015. Project website: https://www.ivir.nl/projects/copyright-in-an-age-of-access-alternatives-to-copyright-enforcement/ (last accessed Mar 2018).

[4] The shortcomings of DRM were precisely and comically skewered by Cory Doctorow in 2004, in a talk at Microsoft Research. As he points out, a dubious aspect of DRM and anti-circumvention technology is that it enables companies to impose new restrictions on the use of information which have no basis in copyright law. http://craphound.com/msftdrm.txt

watermarks, paywalls around newspapers and complex login procedures for the Oxford English Dictionary. Self-sabotaging technology, made to prevent the very thing that the digital world does supremely well – costless copying – is now an industry worth hundreds of millions of dollars, and people are spending entire careers contriving ever-more byzantine means to frustrate and infuriate the rest of us. It's a superb example of mankind wasting its time. In an Open system these costs would be much reduced or entirely eliminated. (Some people will say that these costs pay for the jobs of technologists, bureaucrats and lawyers, but if these jobs don't need to be done, they are unproductive, and freeing these people to work productively can itself benefit the economy).

Moreover, the Amsterdam figures for the Netherlands probably understate the benefits because they focus solely on access and do not consider potential benefits related to creativity and cultural freedom. With unfettered access to what has been done before, there is more to inspire potential artists – more material for them to build upon – and society is enriched.

Reuse is frequent and important in music. Performers play works composed by others, and composers borrow and elaborate the work of previous artists, increasingly directly these days with the growth of sampling in genres such as hip-hop. Reuse fits naturally within the Open framework. As we have seen, anyone would be free to build upon the work of others, but would then be liable to pay a proportion of their own remuneration rights payments (or other revenues) to those whose work they reused.

The major difference from today's copyright regime would be that reuse would be easier and more fluid, because rather than a monopoly right, music labels and creators would have remuneration rights. The moral rights that copyright provides for credit and recognition (or attribution) would be retained in this model, so that artists would continue to have the right to have their work credited wherever it was used or reused.

This new remuneration rights approach could also mean direct benefits in terms of more resources to fund new recordings, which means more music to enjoy. This would depend on how the additional €600 million in value was divided. One option would be to keep payments to music labels and creators at an inflation-adjusted figure equivalent to the current €150m a year, and to channel the additional €600m in value to users. This would mean Dutch citizens paying less than €1 a month but gaining €35 a month in value.

An alternative would be to allocate some of this additional value to record labels and artists. For example, if people were prepared to pay a fee of €3 a month (and some might like to contribute more voluntarily), the money going to artists and record labels could be increased more than three times – a huge amount – whilst consumers would still receive more than €30 of extra benefit. So the Open music model could have huge benefits *both* for citizens *and* for artists and record labels: more artists being paid to make more music that anyone can listen to, at any time, in any way they want.

7.1   Do we need an Open model? Isn't Spotify sufficient?

But do we really need government to put in place a levy, or can we leave private companies like Spotify to sort this out?

Spotify is quasi-Open, and consumers evidently like its flat-fee, unlimited-access model. Yet it has its drawbacks, such as a developing monopoly in commercial hands and the lack of universal access. Moreover, it will *never* be in Spotify's commercial interest to price at a level that gives access to everyone. It is therefore offering a more limited service to both artists and listeners than an Open system would.

Spotify will never provide access to everyone because users differ in their willingness to pay for the service. Some people would pay a lot for it, because they value it very highly – or simply have lots of money. Others are happy to pay a modest amount, and some can or will only pay a little either because music doesn't matter much to them or they don't have the cash. Spotify, however, struggles to tell which is which and has to set a single price for everyone. And a for-profit company like Spotify will never set a price that maximizes access for users. This is because at some point, though lowering the price would bring Spotify more customers, it would not bring enough of them to compensate for the fall in income from each of the existing subscribers. Thus, a private monopoly almost guarantees exclusion of a substantial number of potential users.[5] The Open music model solves this problem by giving universal service and allowing a thousand services to flourish.

At root, platform owners have different interests to the users and artists: open platforms with competition may be great for users and artists, but a monopoly platform is more attractive to investors. Furthermore, a platform like Spotify has strong incentives to use its power to shape the development of the ecosystem in ways that preserve and enhance its grip. In particular, it will want to restrict or kill off innovations or developments that threaten its monopoly – a monopoly which if unchecked will give it power not only over music-listening but over artists and record labels, and over technological innovation related to music access and discovery.

The Open model would not be a proprietary platform taking advantage of artists and users, but a neutral platform democratically overseen, mediating between users and suppliers, setting the rules, levying the charges and setting them to optimize the outcome for society as a whole, including users and creators.

There's no need for the government to set up its own Spotify. There is no need for the state to operate a streaming service or create apps for your phone or store your playlists. All it would do

[5]The same is true on the supply side of the market, where Spotify pays for the recordings it streams. If it had a monopoly, it would be in an even more powerful position when negotiating with record labels and artists, and would push down prices. This would mean fewer recordings being made, to the general detriment. Record labels and artists already worry about the power of platforms like Spotify, and a monopoly would multiply that power. Instead, producers should support an Open music model, ensuring that artists have more outlets, more exposure, and a better deal with more bargaining power in the long term.

is establish a standardized, automatic, blanket-licensing regime, under which any firm could set up a service to provide music by offering the playlists and the apps. Unlike Spotify today, these providers would be mediating as technical distributors only, not as legal distributors. They would not negotiate licensing. In other words, in the Open music model what the state would provide is a universal legal protocol for the licensing of music. Not only would this provide universal access to music and a better deal for artists it would also enable innovation in the search for new technical ways to deliver music to users. This Open legal protocol would create a competitive market of music service providers in just the way that our Open internet protocols have created a competitive market of internet service providers (the companies we all rely on to connect us).

This is not a programme of nationalization. State-sponsored monopolies can be terrible, combining the disadvantages of a private monopoly with an added strata of bureaucracy. In this case, the state's role would be that of promoting competition. It was states which granted copyright monopolies in the first place; if they now introduce blanket compulsory licences instead, and in the process they will be *diminishing* those undesirable monopolies.

Anyone would be free to offer a new recording through any channel they wished and earn a share of the remuneration fund proportionate to the number of plays of the music. Reciprocally, the citizens would all be able to access music any time, anywhere from anyone. Finally, and perhaps most importantly, there would be no restrictions to prevent building upon the licensing platform: anyone could create a new business or a new *kind* of business related to music. Not only would this stimulate innovation in tech industries directly concerned with writing, making and distributing music, it would have an indirect impact on quite different industries, such as restaurants that play music or artificial intelligence startups that need music databases for their learning algorithms.

The same applies equally to other media. Netflix would make just as good an example. Like Spotify, it provides a fixed-price, all-you-can-eat *smorgasbord* of film and television. Like Spotify, it is a platform, with a huge stock-market valuation based on its potential to become overwhelmingly dominant. Like Spotify, it demonstrates how an Open model could work, in this case for film and television. Just as in music there are collecting societies, in television there are practical examples of how a state-coordinated Open model can work, albeit in slightly different form from the remuneration rights approach proposed here. The BBC makes its content freely available to UK citizens, almost all of whom pay a levy in the form of a TV license, and similar models for public service broadcasters exist in many other countries (for example Germany's Rundfunkbeitrag). Yet as with music, the current industry structure for film and television is inefficient, messy and prone to dominance by a single firm. But film and television are an order of magnitude larger than music as businesses, and the benefits we might reap from an Open model are accordingly greater, running every year into billions of pounds (or euros or dollars) of increased value.

Initial funding for music in an Open world could begin at current levels. Total revenues for the worldwide music industry are around $15 billion, but estimates suggest only about 13% of this goes to artists, whereas roughly a quarter is invested in marketing and promotion. In an Open world, artists would generally receive a higher percentage but pay towards marketing of their own music. Suppose, for instance, that 40% of today's revenues were to go to artists. The funding necessary to replace income from monopoly rights would then be around $6 billion a year globally, with the lion's share coming from the US and EU (each around a third of the global music market) and from Japan (18%).

This revenue could be raised by governments by several methods. If done through general taxation, it would amount to less than a dollar a month per person in the US. Alternatively, for example, there could be a levy on digital devices that play or store music. With 264 million internet and mobile data plans in the US in 2015, the addition to each bill would be $0.62 per month.

A more novel approach would be to tax online advertising. Many major online businesses, especially those that use content are heavily funded by advertising. For example, Google, which makes almost all of its money from advertising, relies heavily on the use of content freely from others. YouTube consists of videos and music provided by others, and its search engine would be of little value without access to all of the content on the web, via Google. Similarly, Facebook's revenue comes almost entirely from advertising, but its audience is attracted by content from the users themselves. At present, only a tiny proportion of all these monies are paid to the creators; almost none of Google's search engine advertising revenue reaches them, and even YouTube, which has a revenue-sharing agreement, pays only a small proportion of its revenue to rights-holders. So a tax on online advertising revenues to fund Open information goods is attractive on grounds of fairness and as a transparent means of benefiting artists and rights-holders.

8
How the Secret of Life Almost Stayed Secret

26 June 2000. In Washington and London Bill Clinton and Tony Blair announce the release of the first complete draft of the human genome – our shared genetic code. The achievement is compared to the moon landings or even the invention of the wheel. Missing from the announcement, and much of the coverage, was one key fact. That the genome – nature's ultimate database – would be "open", publicly and freely available for anyone to look at and use, be they researcher, startup company or school-child.

Nor did the coverage make clear how close a call it had been, how very near we had come to having a "closed" genome, controlled and owned by a single private company who would have limited access to those who paid – and would agree to keep the information closed so as to preserve the monopoly.

And this matters: the human genome is a 3 billion long string of letters that provide the recipe for how to make a human being, from proteins to cells to an entire living, breathing being. The database has immense value to science and medicine, it enables us to locate and understand genes and may hold the key to treatments for everything from cystic fibrosis to cancer.

The genome being "open" means that as many minds as possible can set about the work of deciphering, analysing, using and improving. Its being Open rather than Closed has accelerated research and stimulated innovation, already saving lives and generating billions of dollars' worth of social and private value. In 2013, economists estimated that opening this information to all has resulted in a 20-30% increase in subsequent research and product development. And beyond any purely economic considerations, it is obviously appropriate that our common genetic code, shared by every human being, should itself be shared by all. To have allowed it to be owned by a single monopoly provider would have been a travesty.

The human genome is one of the greatest open information projects yet seen. It shows that making information open is both possible and desirable. Public, open information has proved better: it has fewer costs, produces faster results, and has delivered greater value to society. And the genome project was not small: several hundred million dollars of public and charitable funding were needed to make it happen. Anyone who has doubts that open information can be created effectively and efficiently at scale need only look at this example for assurance.

The open genome is also a microcosm for all publicly funded science and research, a cumulative enterprise that stretches from the Royal Society in 17th-century London to the present-day National Institute of Health in the US. As a whole, the scientific enterprise is almost certainly the greatest producer of new information – new knowledge – of all time, and openness is at its very heart. Although it has recently suffered some corruption (notably from proprietary publishers and a creeping pressure to commercialise), publicly funded science remains dedicated to the open creation and sharing of information. The very essence of science is "publication": the sharing of the results of research with other scientists and the community at large. Openness is central to the cumulative and collaborative nature of science. As Isaac Newton knew, each scientist stands on the shoulders of those who went before.

8.1   *The Secret of Life*

On 28 February 1953 Francis Crick stood up in the Eagle pub on Benet Street in Cambridge and shouted "we've found the secret of life". Whether any of his fellow drinkers that lunchtime knew what he meant, history does not relate. But today we do: he meant that he and his colleague James Watson, had discovered the structure of DNA, the substance within the cells of all animals, including humans, that carries the genetic code.

This is the coding for every cell in our body, determining everything from the colour of our hair to the functioning of our kidneys. Watson and Crick had shown that it takes the form of two long sequences of four bases or "letters" – A, C, G, T – woven together in a double helix. (The sequence of 3 billion bases is so long that if you scaled up DNA to the width of a cotton thread, it would extend for nearly 200km.) Within this sequence there are much shorter strings of letters that form the recipe for each protein, and these substrings are called "genes". Within the DNA sequence, there are millions of genes, separated by chunks of "junk" DNA – junk because it does not do anything (or, at least, we don't yet know what it does). Genes encode every protein we need and from the proteins they build all the cells in our bodies.

The discovery of the structure of DNA, in the Cavendish Laboratory in Cambridge, opened vast new horizons for the world. Watson and Crick knew that to work out the full sequence of DNA would mean having the full code of life itself. That sequence, and a map of the genes within it, would be the basis for understanding aspects of biology and medicine as crucial as the basic nature of evolution and the sources of genetic diseases. It would enable scientists to study the mechanisms of cell differentiation, and discover how it is that cells which all contain the same DNA can specialise to do different tasks within the body. For example some become liver cells and clean the blood, while others become part of the eye and enable us to see. If we can understand this differentiation we will be able to work out how to grow new cells of any kind to heal or replace parts of the body that have become worn out or damaged.

So from the moment Watson and Crick announced their discovery in a paper in Nature in 1953, the chase was on to sequence our DNA. However, scientists knew it would not be straightforward. Though the structure had been guessed, there was still no way to examine or read the individual letters. Moreover, even if the sequence had been known, how DNA functioned and how genes actually create a living organism remained mysteries.

Publication of Watson and Crick's paper told scientists that sequencing would be possible, but it would be nearly fifty years before, in June 2000, the International Human Genome Sequence Consortium announced the first working draft of the full human genetic code – and that it would be open to all.

In its final stages, however, the sequencing had become a race between two competing models. On one side were publicly funded scientists committed to producing an open genome available to all. On the other, was Celera Genomics, a private company who sought proprietary control over the genome together with patents over its valuable genes.

8.2 Reading the Code

During the 1970s, the British biochemist Fred Sanger had invented the first practical method of reading the letters in a sequence of DNA. Using this, he and his team had for the first time read the sequence of an entire organism, the bacteriophage phiX174, which they had selected because the entire genome was only 5000 letters long. Their technique involves splitting the DNA into fragments, "cloning" them to create lots of copies, performing complex chemical reactions to sequence the fragments, and then painstakingly joining them back together in sequence. This earned Sanger a share of the Nobel Prize in Chemistry in 1980, and essentially the same technique is still used by sequencers today.

To begin with, each operation was done by hand, and reading a few thousand letters took years, but with the passage of time the technology improved, and by the mid 1980s people were starting to talk about sequencing much larger organisms – even humans. In 1986, at a conference in Santa Fe convened by the US Office of Health and Environmental Research, the Harvard researcher Walter Gilbert estimated that the entire human code could be read for $1 per letter (or $3 billion for the whole genome), and be completed by the mid 2000s. Both estimates were considered optimistic given the state of the technology, and there was significant opposition to spending so much on a single project. However, two important forces were converging, the technological feasibility of the enterprise and a growing interest in biological "big science" by major funders. The human genome was an ambitious, headline-grabbing project that journalists could understand, and the sort of ego-boosting achievement that politicians and billionaires like to be associated with. It was just the kind of project that could attract the vast sums needed to get the job done.

In 1989, Congress approved the funds. Formally established the following year, the Human Genome Project had a target of completion by 2005 and was led by James Watson himself. Other countries were also active, including Japan, France and the UK. In fact the UK, despite its relatively small size was to play a leading role in what followed, largely thanks to John Sulston and his group at the MRC Laboratory for Molecular Biology in Cambridge.

8.3 Of Worms and Men

Sulston is one of those characters whom it would be difficult to make up. He once spent a year and a half looking down a microscope twice a day for four hours at a stretch in order to perform real-time tracking of cell division in the embryos of nematode worms. He is passionate about science as a higher calling – not about fame or money, but as an open and shared enterprise dedicated to developing a better understanding of our world. Humble, dedicated, and replete with glasses, generous beard (and occasionally sandals), he is the textbook other-worldly scientist. Two years after receiving the Nobel Prize in 2002, he was invited to speak in Geneva about openness in science. The organizers were short of funds because of the price of the local hotels, so to help out, Sulston happily volunteered to stay in the local youth hostel. And yet, unworldly though he might look, Sulston has a fierce determination and unsuspected political and managerial capabilities that would prove essential in the years ahead as the genome project grew in size, complexity and urgency.

By 1990, he had dedicated nearly a quarter of a century to studying the genetics and developmental biology of the nematode worm. Crucially, for the past five years he had been creating a "map" of its genome. His group was one of the two leading this field, with the other being led by his close colleague Bob Waterson at Washington University in St Louis. It may seem odd, but this meant they were well positioned to take prominent roles in the race for the human genome.

The humble nematode worm seems a long way from a human. Yet to do something big like the human genome, you need to start small and build up. The nematode worm had been studied intensively for 30 years precisely because it was one of a few perfect "prototype" animals: it was manageably simple, with only 579 cells and a genome of "only" 100 million letters, and yet it was complex enough to have a rudimentary nervous system and behaviours of much more complex animals including humans. Thus, it had become one of a few "stepping stone" species on the path to sequencing the human genome. Not only would its genome be valuable in its own right, but the technology and expertise needed could be directly re-applied to human DNA.

So in 1990, as the Human Genome Project began, Watson invited Sulston and Bob Waterson to take part. At the time, the largest genome ever sequenced was the human cytomegalovirus, which had taken five years to do and had fewer than 300,000 bases. But Sulston, Waterson and Crick quickly agreed the highly ambitious goal of sequencing the worm's three million bases – ten times as many – at a cost of $4.5m in just three years.

They came in ahead of target and on budget. A great part of this achievement was the availability of the new ABI automated sequencing machine made by the company Applied Biosystems. First available in prototype in 1987, this was already orders of magnitude faster than the laborious sequencing by hand. Relative to what came later, the machines were still slow and expensive – several hundred thousand dollars each – but they were the beginning of an advance that has continued to this day. Since 1990 the cost of sequencing per base has dropped a hundred million times, from one million dollars per megabase (million bases) to one cent.

The involvement of Applied Biosytems highlighted the accelerating interest in genomics shown by commercial, venture-capital backed companies. Aided by the growing willingness of the US patent office and courts to permit patents both for genes and gene-related information, more and more capital flowed in. During the late 1980s, this interest was still fairly muted, and when Walter Gilbert, who had shared the 1980 Nobel Prize with Sanger, set up a private "Genome Corporation" in 1987, his colleagues had been shocked. (The company perished in the stock market crash of the same year.) Soon after the Human Genome Project began in 1990, though, gene mapping and sequencing experienced something of a gold-rush, and by 1992-93, activity was growing rapidly.

Most notable was intervention in 1992 of Craig Venter. For the previous decade he had worked at a research institute funded by the American government's National Institute of Health. There he had been an early user of automated sequencing machines. A lot of his work had focused on sequencing short, 200-base strands of DNA at the end of genes (known as "expressed sequences tags"). In 1991 he had strongly advocated patenting of these. This idea, however, had led to conflict within the Institute of Health – and with Jim Watson, who strongly opposed patents on low-level genetic information because of their impact on research and access.

Ambitious and frustrated with what he felt was a lack of support for his approach within the NIH, Venter quit in 1992 and, with $70m of venture capital from an investment company run by Wallace Steinberg, set up of The Institute for Genomic Research. This intended to pursue an alternative approach to genome sequencing, focusing on expressed sequences tags and what were termed "shotgun" methods. While Venter's was a non-profit institute and he and his team planned to publish their work, there was a catch: Steinberg wasn't investing out of philanthropy. In parallel to Venter's institute, Steinberg would establish a commercial company, Human Genome Sciences, in which Venter and colleagues would hold shares and which would have exclusive access to all the data (including the precious expressed sequences tags) for six months, extendable to twelve months if the information looked valuable. Even after that, while academic scientists could see the data, Human Genome Sciences would retain "reach-through" rights to any further commercial developments. The company was an almost immediate success: doing a deal to sell special access to the data to the pharmaceutical giant SmithKline Beecham for $125 million and making Venter and his backers millionaires overnight.

For the public researchers like Sulston, this approach was of grave concern. With its high technology needs, sequencing the whole human genome would cost billions of dollars to complete. With their deep pockets, commercial interests might well be able to outspend publicly funded efforts, and so make the genome data private, locking it up and limiting or denying access to others. Sulston and Waterson, as the most successful sequencers in the world, had already been courted by private firms, and now they saw commercial interests in the US applying increasing political pressure for public funding to be cut off, so that the field would be left free to private for-profit concerns.

8.4   A Wellcome Arrival

At this crucial moment, an unexpected white knight appeared. It took the form of the Wellcome Trust. The Trust was formed in 1936 on the death of Sir Henry Wellcome, a pharmaceutical entrepreneur and tycoon. Wellcome bequeathed the Trust all of his shares in his company Wellcome Foundation, later Wellcome plc, then Glaxo Wellcome which was finally absorbed into GlaxoSmithKline. Initially, the Trust had grown fairly slowly, but it benefited enormously from the rise in the stock market in the 1980s, which had been especially steep for pharmaceutical companies. Then it had received a particular boost from the success of anti-HIV drug AZT, which was owned by Glaxo Wellcome. In 1992 the Trust had sold some of its shares and become the wealthiest private medical research charity in the world. Its annual budget doubled from £100m to £200m a year ($350m in 1992 and nearly $1bn in today's money).

This could not have happened at a better time, since 1992 was the crunch year for the Human Genome Project. Not only was there the growing threat from the proprietorial commercial players, but the end of the project's initial three-year grant was looming in 1993. The issue was especially acute for Sulston and his team at the Laboratory for Molecular Biology in Cambridge. They were powerfully aware that they were already stretching the limited funds available from the Medical Research Council, whose budget was a fraction of the billions that the National Institute Health in the US had at its disposal. But suddenly, the UK had a funder with deep pockets, and it was private, which mattered for two crucial reasons: it was (largely) immune to political pressure and it could move much faster than any public-sector funder with all the constraints of reviews, checks and government approval.

Sulston went to Wellcome for funding. In just a few months the Trust reached a decision, and in the summer of 1992 it agreed to commit £40-50m over five years to his group, not only to continue its work on the worm, but to accelerate its work on the human genome. Included in the grant was support to create an entirely new campus just outside Cambridge, to be called the Sanger Center.

Even better, the Wellcome grant spurred action in other quarters. Put on its mettle by the Wellcome grant, Britain's Medical Research Council responded in June 1993 by committing more than it ever had before: £10m over five years to complete the worm sequence. Just three years before, there was funding only to do 3% of the worm genome over three years; now the target was to complete the remaining 97% in the next five years and work on the human genome at the same time. Wellcome's grant spurred action in the US, where the National Institute of Health also increased its funding to Bob Waterson's lab.

With this new funding, the whole process scaled up and became industrialised. Sanger soon had more than 200 people running dozens of sequencing machines twenty-four hours a day, seven days a week. Ambitions grew as to what was possible.

And as the worm wriggled forward and technology improved, attention began to turn to the human genome – the holy grail, but which was thirty times larger and more expensive than the worm's, and much more complex because of the greater proportion of repetitions in its sequence.

Still uncertain as to what could be achieved and in what timescale, Waterson flew to Cambridge to visit Sulston in the autumn of 1994. On his way home, he came up with what he termed an "indecent proposal": to escalate human genome sequencing to 600 megabases a year, with the effort split three ways between his own lab, Sulston's and a third to be identified. At this scale they would be able to sequence at 10 cents per base (a tenth of the cost estimated a decade earlier) and complete the sequence by 2001 for $300m.

It was a highly ambitious proposal. At this point, less than 1% of the human genome was sequenced and most of that consisted of small fragments. Sequencing rates were less than a tenth of what Waterson was proposing. And who would put up the money? The proposal needed $60m a year for five years – an unheard of sum in biology for a single project.

The proposal was also exciting, and a blueprint for what was to come. But funders and the research community were disappointingly slow and further disappointments were to come. In 1996, Sulston finally submitted a new bid for funds jointly to the Medical Research Council and Wellcome: £147m over seven years to complete his third of the genome. Unfortunately, the MRC just did not have the money. Although Wellcome agreed to provide half – £60m – the MRC would only continue its support of £2m a year for five years.

With this level of funding Sulston would only be able to do half of what he wanted: one sixth rather than a third of the genome. In the US, Bob Waterson had a similar experience, being awarded only a quarter of what he asked for the next two years. The plan was that the National Institute of Health would re-evaluate progress in 1998, and make new grants only then. In part, this was the result of politics and disagreements within the research community.

8.5   The Risks of Delay: the Saga of BRCA2

As a warning about the risks of any delay, scientists needed to look no further than the saga then playing out around breast cancer genes. In December 1990 Mary-Claire King at University of Berkeley California had identified a mutation of chromosome 17 (BRCA1) that was associated with a high risk of breast cancer. Then, in the summer of 1994, Mike Stratton, at the University of Surrey in the UK, identified a similar mutation, on chromosome 13 (BRCA2), which was also associated with a high risk of breast cancer.

Knowing that Mormons kept excellent genealogical records, which are a great resource when searching for genetically inherited disease risks, Stratton had been working with a colleague, Mark Skolnick, at the University of Utah. Skolnick had set up a private company called Myriad Genetics specifically to look for cancer genes and then patent the genes and any associated tests they could design. Stratton was aware of the existence of Myriad and shortly before he located BRCA2 he asked Skolnick what would happen if they did locate and clone it. Skolnick told him that Myriad would patent it. But Stratton was deeply concerned. As he stated later, it became clear to him that there was a clear risk of a "conflict between the clinical and ethical imperatives and the commercial imperatives" – between ensuring that patients got the benefits of new tests and treatments derived from this knowledge, and restricting access so as to charge high fees for any tests or treatments. "Myriad had a duty to service the needs of investors", he said. "I realized I would have no influence on how the discovery was used."

Stratton ended his collaboration with Skolnick immediately after identifying the location of the gene, but now he found himself racing the Utah lab to identify precisely the gene and clone it, because "locating" a gene is not the same as being able to isolate, clone and then sequence it, all of which are necessary to qualify for formal publication or for a patent application. Stratton quickly enlisted the help of John Sulston and the Sanger Institute, and by November 1995 they had the sequence and rushed to publish in Nature so that the data would be in the public domain and unpatentable by Skolnick.

Unfortunately, despite Stratton's efforts to keep the information secret even from close collaborators, somehow enough leaked out for Skolnick to be able to help his own team identify the gene and submit their patent application – just one day before Stratton's paper came out in Nature on 28 December 1995. Despite Stratton's efforts to fight back, Myriad now claimed patents on both BRCA1 and BRCA2, for although Skolnick had not discovered BRCA1, his lab had been the first to clone it.

Successful in its applications, Myriad set up a lab to perform tests, charging $2,500 per patient, and moved aggressively with legal suits and threats against anyone else who sought to offer tests more cheaply. Myriad restricted other labs to doing simpler, less effective tests, for which they had to buy a license for several hundred dollars per patient.

As Stratton bitterly said of the BRCA2 test, "Myriad is claiming a fee from all women who undergo tests in the United States for a mutation that was discovered by us." Stratton felt he now had no choice but to try to secure some patents of his own – though only to use them defensively to fight back against Myriad. These patents did turn out to be useful, but unfortunately only in Europe. For example, Britain's NHS refused to license from Myriad and instead carried out the tests itself using Stratton's publication and patents. In France, the Institut Curie, with the backing of the French Government in September 2001, launched a formal objection to Myriad's patent, and after a battle lasting several years the patents were struck down. Despite these victories, however, Myriad was successful in the larger battle: worth over $3bn in 2015, it has made its founders very rich.

8.6   Back to the Genome

For Sulston and others in the public research community, the message was clear. Almost every aspect of the discovery of the BRCA genes had been publicly funded. Even Skolnick's valuable

Mormon genealogical database had been largely funded by public monies when it started in the 1970s. Yet, a well-funded proprietary firm had jumped in at the last minute and claimed broad monopoly rights that would not only limit future research but would, through high prices and other restrictions, deny life-saving diagnostics to patients around the world. This example of private profiteering showed that it was essential to get the full human genome into the public domain as soon as possible, so that no one would be able to claim a similarly broad monopoly of low-level genetic sequences.

Venter had not gone away, and soon the worst fears of Sulston and his colleagues were realised. In 1998, the very year the public consortium planned to accelerate its efforts, Venter launched a new company: Celera Genomics. This had a single aim: to sequence the human genome and exploit it financially. Launched without warning at a major press blitz on 10 May 1998, just two days before the annual meeting of the Human Genome Project, its message was clear: the race was on. And Celera was a serious competitor, backed by $300m from Applied Biosystems, the producers of the sequencing machines themselves. Moreover, as a commercial venture, its stance was clear: not only would it pursue patents on any genes it identified, but the sequence would be released – if at all – only after a delay and probably with restrictions on use. The prospect of an open genome was facing the greatest of threats.

Venter and his colleagues skillfully played on the American aversion to funding public efforts that might compete with private enterprise – an aversion that was especially strong in a Congress controlled by Newt Gingrich's Republicans. Venter and Celera had given an exclusive pre-briefing to Nicholas Wade of the New York Times, whose piece on the day of the launch stated: "Congress might ask why it should continue to finance the human genome project through the National Institute of Health . . . if the new company is going to finish first." This was a doubly brilliant attack on the public project not only insinuating as fact that Celera would finish first but that public funding was unnecessary and unjustified. Much more of the same was to come in the succeeding two years, with repeated claims that the private project would beat the public one. As Sulston ruefully observed years later: "Craig was no longer in science, he was in business. And the priority for a business is not scientific credibility but share price and market penetration. Trying to get reporters to report the admittedly more complex analyses . . . would be an uphill battle. We were learning fast that we would have to play the public relations game if we were to survive."

Would funders lose faith and cancel their support for the public effort, or (almost as bad) fail to increase their support sufficiently for the public effort to properly compete? At this crucial moment, the Wellcome Trust once again came to the rescue. By chance, on the Wednesday after Celera's announcement, Sulston had a meeting scheduled with the Wellcome Trust to seek more funding. After an emotional plea for the importance of free genomic data, he waited anxiously outside for the Trust's decision. It was not long in coming: unanimous approval for a doubling in funding to £120m and full support for Sanger to take on a third of the genome by 2001.

In many ways, Venter and Celera's aggressive approach backfired, as the Wellcome program officer reported: "Once the governors realized that Craig Venter's initiative was essentially a privatization of the genome . . . there was no risk they would pull out." By the day of the meeting, he added, "everyone's dander was up . . . The governors just said: 'We must do this.'"

It was a huge vote of confidence. Sulston and Michael Morgan of the Wellcome Trust immediately got on a plane to fly to Cold Spring Harbor to announce the good news to the Human Genome Project conference.

On the Friday morning at Cold Spring, before a packed crowd, Sulston and Morgan announced the Wellcome Trust's doubling of funding and its commitment to a public, free, open genome. To clarify the Trust's motivation and commitment, Morgan added that it was opposed to the patenting of basic genomic information and would fight such applications in the courts. The room erupted. Everyone knew what this meant. Rather than the Human Genome Project being dead – as many had feared on the Monday – the project was suddenly stronger than ever. The pressure would now be on the US National Institute of Health to match the Wellcome's initiative, and it did, committing a further $81.6m in 1999 (plus $40m from the Department of Energy). The threat had brought together this transatlantic community as never before.

The next two years would pass in a rush. Less than a year later, in February 1999, the leaders of the Human Genome Project met in Houston and committed themselves to producing a "working draft" of the genome by mid 2000, a year earlier than they had planned just six months earlier. The teams were working at full pace. In 1999, Sanger aimed to sequence three times as much as in 1998 – and more than in the whole of the seven years before that.

On 2 December 1999, the Human Genome Project announced one of its first major successes: the publication in Nature of the complete sequence of an entire chromosome – chromosome 22. The announcement included the discovery of 545 genes, more than half of which were previously unknown. Chromosome 22 was also implicated in over thirty-five diseases, including some forms of heart disease and leukaemia, so this was of major benefit to medicine. By publishing the information openly, the Human Genome Project ensured that none of the genes could subsequently be patented, and that researchers in either the public or private sector were free to start using it immediately, with no need for licenses or risk of lawsuits.

But Celera showed no signs of easing off. In a tense conference call on 29 December 1999, there was a showdown between the two sides. Taking part from Celera were Venter, Tony White (CEO of Applied Biosytems) and three other executives; on the public side, Sulston and Waterson were joined by the director of the National Institute of Health, Nobel Prize winner Harold Varmus and his colleague Francis Collins, and the Wellcome Trust's Martin Brobrow. During the meeting it became starkly clear that Celera had no intention of publicly releasing data despite their stated commitments on this front. Tony White of Celera wanted any joint database to exclude commercial competitors for 3-5 years – an eternity in this fast moving field, and wanted the publicly funded consortium to stop work as soon as there was a complete draft. This would have meant the publicly funded effort doing most of the work and then stepping aside, leaving Celera control with exclusive advance access to the joint database and all of the immense commercial and scientific possibilities it opened up. It was an aggressive demand and the public team flatly refused.

By now, any real hope of collaboration was dead, though Francis Collins of the National Institute of Health, under immense political pressure to collaborate with the private sector, would continue to do all he could to build bridges. In March 2000, after news leaked of Celera's refusal to engage in meaningful collaboration, the Washington Post said the project had become a "mud-wrestling match".

Despite all the heat, there was some light. In response to the debate, Bill Clinton and Tony Blair issued a joint statement that the human genome sequence should be freely available, at least to researchers. Though the race was now nearing an end, this was valuable affirmation of the public, open approach and helped close out any risk of last minute compromise.

Nevertheless, with an election looming, Clinton was anxious to resolve what could be a damaging public-private conflict. So, despite hugely unequal contributions, when the moment came to make the announcement on 26 June 2000, it was claimed a joint victory by Celera and the Human Genome Project. In the White House, Clinton was flanked by Venter on one side and Francis Collins on the other in a symbolic show of unity.

It only remained to compare the work of the two groups. Celera had always had the benefit of full access to the public team's data, which was released daily, whereas Celera had not released its own data at all (despite initial promises to release it quarterly). By the time of the White House announcement, each claimed to have a complete working draft of the human genome. Joint papers were planned for February 2001 in the journal Science. Each side would publish its sequence for the first time. But before publication, a controversy broke out as it became clear that Celera would not release its data into a public database. This is a standard requirement for publication in this kind of journal, since it is essential to allow for full scrutiny and reuse by the academic community. Nonetheless, in the face of vociferous protests, Science agreed to bend its rules, whereupon participants in the Human Genome Project unanimously agreed to withdraw from Science and instead publish in Nature.

Finally, a few days before the official release of the papers, on Monday 12 February 2001, the two sides exchanged their papers. For the first time the public team could read Celera's results. The experience was shocking. As John Sulston wrote: "We had fully expected their sequence to be better than ours, given that they had access to all our data and we knew they were using it. But they were publishing a sequence that seemed overall no better than the publicly released sequence, and which depended heavily on it." He believed that Celera would have had no draft genome at all without the public project, and concluded that its "chances of **ever** having a fully finished sequence would have been very slim indeed."

Further analysis bore this out. A year later, one of the world's leading experts wrote that Celera had been dependent on the public data in three different ways, and that "even with the public data, what Celera calls whole-genome assembly was a failure by any reasonable standard: 20% of the genome is either missing altogether or is in the forms of 116,000 small islands of sequence that are unplaced, and for practical purposes, unplaceable."

Two myths remain: first, that Celera's sequence was more cost-effective even though less accurate; second, that the competition from Celera was ultimately beneficial for science – for example, because it caused those "go-slow" scientists to pick up their pace. However, neither claim is correct: later analysis showed that none of the savings that Celera's methods were intended to deliver were realised, and that the duplication of work meant that the overall costs were significantly higher than necessary. And on the public side, the need to compete meant going faster than would have been optimal, leading to some major inefficiencies which eventually increased the total cost. When the publicly funded team had to shift to producing a "draft sequence" of reduced quality, it postponed the production of a higher quality sequence – and may even have been putting its eventual achievement at risk, because funding might have fallen away once the project was announced as "complete".

Ultimately, we cannot know precisely the impact of this scientific race. But what we do know is that the public team produced a far higher quality sequence – indeed really the only sequence – at approximately the same cost as the private project. Most importantly, they produced a sequence that was publicly available to all researchers, private, academic or commercial.

Almost from start to finish, the story of the discovery and sequencing of the human genome was one of openness and public funding. An open genome is one of mankind's great scientific achievements, and provides a basis for future research and innovation. And it might not have been so.

We should leave the last word to John Sulston: "Deciphering the information will take a long time and need every available mind on the job. And so it is essential that the sequence is available to the whole biological community ... When the commercial company that became Celera Genomics was launched ... the whole future of biology came under threat. For one company was bidding for monopoly control of access to the most fundamental information about humanity, information that is – or should be – our common heritage." He paid tribute to the public bodies funding the Human Genome Project for deciding not to leave the field to Celera, so that today any scientist anywhere can access the sequence freely and use the information to make his or her own further discoveries. But as Sulston wrote, we should remember "how close we came to losing that freedom."

9
Meet Jamie Love

The establishment of the World Trade Organization in 1994 was the most significant trade agreement of the 20th century. It had three main parts. Two were classic trade agreements that sought to remove barriers to trade, both in traditional physical goods (GATT) and in services (GATS). But the third agreement, TRIPS, is quite different. TRIPS stands for Trade-Related aspects of Intellectual Property Rights, but strangely this treaty said little about trade. It was all about intellectual property rights – and the expansion and enhancement thereof.

The justification offered for the inclusion of TRIPS was that by signing the other two trade agreements, developed countries, and especially the US and the EU, would be opening up their markets to competition from lower-cost developing countries. Although the developed countries would benefit from cheaper goods, the change would have even more benefit for developing countries. In return, developed countries wanted more. Their information-based industries – from software to pharmaceuticals – were a large and growing section of their economies, and whereas physical goods can be stopped at a border, it is almost impossible to erect technical barriers to the flow of that information. Once a piece of software is published, it can be copied by anyone; if the recipe for a drug is published in one country, it can easily be copied in another. So the developed countries wanted more "protection" for their information industries. The third component of this huge free trade agreement therefore was a treaty dedicated to *reducing* the flow of information. As a result of TRIPS, for instance, India had to introduce product patents for pharmaceuticals, and the United States modified and extended its copyright code.

Special interests such as the pharmaceutical lobby in the United States had played a key role in drafting and promoting TRIPS as part of the WTO package. The impact of this agreement was immense, not only economically but on the framework of global information, yet there was almost no public discussion. How can this be?

The challenges of the modern world are exacerbated by their complexity and interconnections. Human beings struggle with this. We want simple problems with simple solutions. But the complexity that confronts us and the rules that govern our world – and which shape and shift power – are practically opaque. This applies not only to us, but to our representatives as well. Generalists in a sea of detail, they have no way of knowing, much of the time, what the effects will be of what they are doing.

I remember watching the final vote on the Software Patents directive in the European Parliament in 2005. On a voting day there, each party produces a "voting list" instructing its MEPs how to vote (and why) on each amendment that will come up (usually there is not a single text but a proposed text and then a bunch of amendments, each to be voted on). On this occasion the voting list was over six hundred pages, and that was just one legislative session. As you can imagine, most MEPs have no idea what they are voting on and the ordinary citizens aren't even aware that they *are* voting. This is not necessarily opacity by design (though undoubtedly it can be useful and is manipulated for commercial and political ends). The complexity of our legislation reflects the complexity of our world. After the financial crisis of 2008-9 the US government responded with the Dodd-Frank Act. Not a single person on earth understood it, or even read it, in full, since it runs to nearly a thousand pages, with associated regulations that now exceed 13,000 pages in length.

Unfortunately, complexity creates inequalities in power, above all the power to influence the rules that run societies – rules which in turn beget more power and influence. Concentrated interests, usually corporate but also very wealthy individuals, are better able to handle complexity than the rest of us. Not only do they have the resources to buy teams to understand the issues and present a case, but they are better able to extend power through time and space: to send lobbyists hundreds or thousands of miles to seats of power, and to pursue their interests persistently.

Complexity – and the interconnectedness that is part of it – has helped to centralize power ever further. Today, if you live in the EU, it is likely that more than half of the laws and regulations being made for you are from Brussels, not your national government, and meanwhile domestically over the past fifty years, power has almost certainly shifted from local to the central government. This is true all round the world, from Indonesia to Brazil.

Such centralization has the unintended consequence of moving rule-making away from voters both literally and metaphorically, making it easier for special interests to exert their influence. It is easier for a corporation to have one large office of lobbyists in Washington DC than an office in each of the fifty States of the Union. For ordinary citizens, though, the opposite is true: keeping up with and influencing decisions is much easier when they are made on your doorstep.

Politicians and bureaucrats are charged to preserve and pursue the public interest on behalf of the electorate, but they struggle to handle the growing complexity of the modern world, and find it increasingly hard to resist outside interests even if they wish to. This a fundamental challenge for modern democracy; but this bigger question is not my focus, rather it is on the relation of this mind-boggling complexity to the regulation of information in the digital age.

Digital technology is complex, fast-moving and fundamentally abstract. Unlike a child going hungry or a park being built over, digital policy is not a visible or popular political cause. The issues involved require value judgments, such as who should own and control the cables that transport the bits the internet relies upon, or changes in copyright law which subtly but significantly change the distribution of money and influence between media conglomerates, artists and the general public. And because digital information involves enormous international networks, regulations are decided on a supra-national basis, often in remote and privileged locations far from everyday political processes, such as Geneva. The result is to reduce external scrutiny and to leave lawmaking to technocrats and corporate lobbyists with their special interests.

Blue eyes, a kind face, a strong American accent. Until Jamie Love speaks you can easily imagine him as an academic or bureaucrat – or even, given his smart appearance, a corporate executive. But once you hear him speak, that changes. A passion, an anger even, steams off him, and is evident, along with an incisive intelligence, in every word he says. It's a passion forged by working for years in obscurity on difficult issues in the front line of the "information wars".

Starting in the mid-1990s, almost entirely alone, he made pilgrimages abroad to intervene in the meetings and deals where the digital future was being carved up. Long before almost anyone else, he realized that this was a special moment, when key rules of the information age were being made. Almost no one else was watching or reporting what was going on. Gradually, the fuss he made, the speeches he gave and the articles he wrote alerted others, and there is now a much wider appreciation of the issues, though there is also more to fight for.

I first met Jamie in September 2004. Over the previous year or so I had become interested in "information activism" and I was just acquainting myself with the area. Somehow I stumbled across an event that was taking place in Geneva. Most of the people making the running seemed to be in the US, so it was exciting to see them appearing in Europe, and the roster of speakers was remarkable. The event was called "The Future of WIPO" and was organized by something called the Consumer Project on Technology (CPT), led by Jamie Love and his wife Manon Ress.

Blank about all this, I had to look up WIPO and CPT. The World Intellectual Property Organization turned out to be a UN agency dedicated entirely to promoting "intellectual property".

From my reading, I already knew that intellectual property wasn't nearly so unabashedly positive as its name suggests. I'd never been to Geneva and associated it with little more than luxury and a lake, but there were cheap air tickets and a youth hostel, so I invited myself. The event was quite small, just fifty or sixty people.

During the later '90s, the World Trade Organization (WTO) had become a household name thanks to anti-globalization protests, but even now WIPO remains unknown to most people. Yet it was at WIPO that many of the rules of the internet age were turned into treaties which the member nations were obliged to enact into law. Acts and directives such as the Digital Millennium Copyright Act in the US and the Copyright Directive in the EU derive directly from a WIPO treaty in 1996. You may never have heard of those acts or directives either, but they have shaped your experience of the digital age, through services such as YouTube, TiVO and more.

Moulding of the structure of the information technology by corporate interests is not new. It happened with the telegraph in the 19th century and with radio and TV in the 20th. What was new in the 1990s was the internet. Connecting the world more intimately than ever before, it was more Open and more democratic than anything before, and not everyone liked this. "Intellectual property" now involved massively higher stakes.

For nearly a decade, Jamie and Manon toiled almost on their own, unknown and barely funded. They had to watch as the lobbyists for the media conglomerates that owned the major labels and studios and outlets rigged the WIPO Copyright Treaty of 1996, which laid out how copyright would function on the internet. But gradually they built awareness and were joined by others, analysing, challenging, objecting.

In such a difficult area, against stacked odds, successes are small and compromised: amendments made here and there, a rule slightly less bad than it would have been, a treaty stalled or even stopped (usually only to be replaced by something only marginally less bad). This is not high drama that makes good news copy or television. It is work in darkness. Year after year of peering through mind-numbing legalese to see to the play of wealth and power beneath. It involves explaining what the wording means and why it should be revised, again and again and again, to officials who are at best mildly sympathetic, at worst in their posts precisely because they're aligned with the prevailing interests.

But Jamie and his evolving coalition did make a difference. Jamie managed to stop a broadcasting treaty proposed by – you guessed it – broadcasters, which would have given them new special monopoly rights in their programmes. In another instance, the coalition persuaded WIPO to add legal exceptions to provide better support for blind people and those with disabilities to access copyrighted work. Alas, part of the result of all this work was to alter the mechanisms by which the rules are drawn up. The US Trade Representative, the real American power in these matters, and almost entirely the creature of the major corporate interests, gradually relocated "intellectual property" and other key information regulation out of international fora such as the WTO or WIPO and into bilateral negotiations where the US and the lobbyists could more easily exert their will.

The sheer strength of that will became clear to me in 2009, when I travelled as an academic to the European Parliament to watch it debate a Directive that would extend copyright in *existing* recordings. This was one of the most blatantly partisan – or even corrupt – of possible changes. To add 20 or 40 further years' copyright to the prevailing 50 was simply to extend the monopoly for back catalogues. It was in effect a shockingly regressive tax on EU citizens for the benefit of a few multinational record companies (Sony, BMG, Universal Music Group) and a few hugely successful artists such as the Beatles, the Rolling Stones and U2. This goes against the very purpose of copyright which is to incentivize and reward creators for making *new* music.[1] There is no way for the Beatles to get in a time machine and record another album in 1965

---

[1] The extension also lengthened copyright for new recordings. However, the incentives for this are so negligible as to be irrelevant: an uncertain gain of small amounts of extra money fifty years or more in the future is economically insignificant both to artists and investors.

because their copyright was extended in 2009.

So I went to meet MEPs to try to persuade them that this was a mistake. The record labels and collecting societies had permanent staff in Brussels who had done a good job of lobbying. I was probably the only person many of the MEPs ever met who would make the case for the other side. Even the Commissioner who introduced the Directive admitted to me that of the 27 meetings he had been to with interested groups, 26 were with groups promoting it (the exception was from the Consumers' Union). When I visited one British Labour MEP who was playing a significant role in establishing the position of his group, he was so incensed that anyone should oppose longer copyrights that he yelled at me and almost ejected me bodily from his office. Afterwards, I discovered that he had worked for many years in the music industry. He was not motivated by anything other than long-ingrained conviction, but he certainly wasn't listening to the merits of the case.

10
Openness: The Best Medicine

Patents and copyrights do, of course, exist for a reason, so let's look at one crucial instance. Patents support the research and development of new medicines. Formulating and testing new drugs is extremely expensive. Without monopoly protection, the argument goes, competition (copying of drugs by other manufacturers) would drive drug prices down so far that pioneers would see little or no return on their substantial investment. Without the anticipation of high return companies and their investors might never risk the expense of research, and rather than high-priced drugs we'd simply have no drugs at all.

This logic is not wrong, so much as misplaced. There are Open-compatible ways to fund the development of new medicines that are more effective than patents at rewarding innovators and stimulating innovation. We can make medicines available to everyone at the cost of manufacture *and* fund medical innovation at, or even above, the level we do today.

In any case, the situation today is hardly satisfactory. In many countries, high prices for medicines are an everyday concern, and even where many people have free prescriptions, they are paying those high prices through their taxes. Americans spent over $400 billion on pharmaceuticals in 2016. That is $1,400 for every man, woman and child, whether ill or not, and this average conceals the everyday reality that individuals who actually fall sick may have to spend tens or hundreds of thousands. Millions of Americans, struggle to afford what they need, and some even will die for lack of medicines. And if citizens in the richest country on earth can't afford medicines, imagine the plight of countries that are much poorer or facing epidemics.

During the AIDS epidemic of the 1990s and 2000s which ravaged Africa and other parts of the world, the principal anti-retroviral treatments were all under patent with big multinational pharmaceutical companies such as Pfizer, GlaxoSmithKline and Boehringer, which kept prices much too high for most patients or for the governments to afford. In South Africa, basic treatment costs were more than two thousand rand ($250) per month in 2002 at a time when per capita GDP was $250 a month. To maintain these prices, the companies refused to allow the manufacture of generic versions. Tens of thousands of people were dying and campaigners were desperate for a change.

In 2002, members of the Treatment Action Campaign filed a complaint with the South African Competition Commission. The lead complainant, Hazel Tau, a single woman from Soweto, wrote in her submission that she was the family breadwinner.

> I was diagnosed with HIV in 1991 ... Since April 2002, I have not been so well. I have had an increasing number of opportunistic infections including ... a lung infection, which was suspected to be pneumonia ... I have also lost a lot of weight. I weighed about 75 kilograms up to about 2000. I have lost over 25 kilograms since then ... I need to go onto treatment given that my CD4 has dropped below 200 ... anti-retroviral treatment is required. But I cannot afford to pay even R2,000 a month for this. If the prices of anti-retrovirals were reduced to between R400 to R500 a month, I could afford treatment on my present salary. I am aware that I will have to sacrifice some things, but I know that this treatment will help me and keep me healthy. I cannot afford to pay the prices the drug companies charge for anti-retroviral treatment.

At this point, Jamie Love and his team at the Consumer Project on Technology re-enter the picture. The impact of the high prices of drugs, especially retrovirals in developing countries, had been a major concern of Jamie's for years, and he was among those who provided expert evidence to the South African Competition Commission on the economic side of the case. In all, he wrote or co-wrote six of the expert opinions.

On 16 October 2003, the Competition Commission ruled against big pharma and in favor of Hazel Tau and the hundreds of thousands of other AIDs sufferers:

> Pharmaceutical firms GlaxoSmithKline South Africa (Pty) Ltd (GSK) and Boehringer Ingelheim (BI) have contravened the Competition Act of 1998. The firms have been found to have abused their dominant positions in their respective anti-retroviral (ARV) markets.

As a result of this decision, major pharmaceutical companies agreed to license their patents to generic manufacturers on reasonable terms – and not only in South Africa but across sub-saharan Africa. Prices dropped immediately and have continued to fall. Between 2000 and 2014, Médecins Sans Frontières estimated that prices fell 99% to around $100, and much of that is due to this victory.

This was an example of the benefits of *removing* patent monopolies, but we also have a striking example of the costs of *introducing* them. For the twenty years before the TRIPS agreement of 1994, India had forbidden patents for pharmaceutical products.[1] However, under the agreement – to which India is a signatory – patent protection must be provided for pharmaceuticals, to allow the holders to raise prices. Until then, without patents, any firm in India could manufacture a given drug: and they did. India had a booming industry in generic medicines – those that can be produced by anyone (often without the fancy brand names), because there are no patents. Drugs were cheap and even poor people could afford them, but in theory this meant less revenue for the original creators. India is therefore a good

---

[1] Strictly, India did not have "product" patents but had "process" patents for pharmaceuticals. Patents on the recipe for a drug did not exist but a company could register a patent on a specific, novel way of manufacturing a drug.

test case for what these trade-offs mean in practice, and thanks to a case study that focused on a major category of anti-bacterials called Quinolones, we have some numbers.[2]

The paper estimated the cost to the nation in just this one segment would be some $350–500 million a year, falling primarily on consumers but also upon local manufacturers. Yet the gain to the owners of the patents was a mere $50 million a year (gains to patent owners can be much lower than the costs to consumers because increased prices mean lower sales and those lost sales are a loss both to consumers who get no medicines and to manufacturers who get no revenues – this is the so-called deadweight cost of economists, in this case a very appropriate term, since lost access to drugs could literally mean death). So the net cost to India of introducing patent monopolies was $300–$450 million a year. And these are just the dollar numbers. Think of the costs in misery, of the people who can no longer afford treatment, whose illnesses are prolonged or whose lives were shortened unnecessarily.

Medical patents, then, can have appalling financial and practical consequences for millions. But what of big pharma's argument that they are a just and necessary reward for the expenses and risks of research? That without patents there would be many fewer medicines and many more who miss out on treatment? That too needs examination, because almost every innovative medicine we now have started with work in a government-funded research lab – and many of them were completed there too. This is especially true of our greatest advances, from Pasteur's germ theory to Fleming's discovery of penicillin and right through to today's work on gene therapies and predictive medicine. Alexander Fleming's work was paid for by public institutions. Accordingly, he did not hide or patent the discovery of penicillin, but published it for everyone to test, use and build upon, so helping to save millions of lives.

On the same principle, current and future work that is paid

---

[2]Shubham Chaudhuri, Pinelopi K. Goldberg, and Panle Gia, *Estimating the Effects of Global Patent Protection in Pharmaceuticals: A Case Study of Quinolones in India,* Yale Working Papers, 2003; repr. *American Economic Review,* 2006.

for collectively by the public is usually Openly available. And this is a huge proportion: almost half of all medical R&D in the world today is funded directly by governments, and in basic scientific research the proportion is much higher. In addition, UNESCO estimated that in 2012 private, non-profit financing of medical R&D in the US amounted to just under $15 billion. This includes both the "mega-philanthrophy" of super-rich individuals and the aggregate contributions of many small donors, such as charities that focus upon particular diseases, and almost all of it is Openly published.[3]

One of the reports that Jamie Love wrote for the Competition Commission in South Africa investigated R&D costs and challenged the assumption that they are borne by the companies that end up with the patents. This is rarely true, he argued, especially for drugs for diseases such as HIV/AIDS. For instance, a patent for one of the most important of the anti-retrovirals, AZT (marketed as Retrovir) was granted to the Burroughs Wellcome company in March 1987, and then acquired in a takeover of Burroughs Wellcome by GlaxoSmithKline. Burroughs had not been shy in claiming primary credit for the development of AZT, but the facts were a good deal more complex.

Though the patent was granted in 1987, the drug had first been synthesized in 1964 by Dr Jerome Horowitz of the Michigan Cancer Foundation, supported by a US government grant. Its use against animal retroviruses was first demonstrated by Wolfram Ostertag at the Max Planck Institute in 1974 using mice, and was

---

[3] Traditional state funding of academic research pays for effort, not specific results. It allows researchers to determine their avenues of study partially or entirely for themselves and is not directly dependent upon the achievement of particular goals. This is particularly useful for funding basic research, which can be very long-term indeed, and high-risk research, which may be worthwhile because of occasional big breakthroughs even though mostly it produces no or negative results. This kind of funding often envisages a combination of research with teaching, which is desirable for both the teachers and the taught, and is itself a form of long-term investment. The results are unpredictable, and very hard to quantify, but can be spectacular, and have included many of the great achievements of every academic discipline.

again supported by government funding (this time not American). Next came crucial clinical research including the first test to see whether AZT was effective against human immuno-viruses such as HIV, and at what specific concentrations it was effective. This too was carried out by government-funded researchers in the US, this time at the National Cancer Institute at Duke University. These researchers, none of whom were funded by Burroughs Wellcome, were also the first to administer AZT to a human being with AIDS, and performed the first clinical pharmacology study in patients.

As some of the key scientists wrote in a letter to the *New York Times* in September 1989, not only was almost all of the development of AZT carried out with public funds, but Burroughs had actually retarded developments in the final stages. The demonstration of clinical effectiveness had, they pointed out, been accomplished

> by the staff of the National Cancer Institute working with staff at Duke University. These scientists did not work for the Burroughs Wellcome Company. They were doing investigator-initiated research, which required resources and reprogramming from other important projects, in response to a public health emergency. Indeed, one of the key obstacles to the development of AZT was that Burroughs Wellcome did not work with live AIDS virus nor wish to receive samples from AIDS patients.

Presented with a working drug, the result of several decades of research, Burroughs Wellcome had merely carried out the final set of clinical trials to win approval for use from the regulator, the US Food and Drug Administration (FDA). And even here, they received assistance. In the US, AZT had been designated an "orphan" drug (one for use in a small patient population), which meant that half the costs of clinical trials would be paid for by the government, through a tax credit to Burroughs.

Under the Open system, Burroughs Wellcome would not have been handed such an enormously valuable patent on the basis of work done, in large proportion, by others, including researchers paid for by taxpayers. Instead, private R&D would be Open in the same way as publicly funded R&D. The science would be available for use by everyone; manufacturing would be unrestricted and competitive, and this would keep prices close to the cost of manufacture, just like generic medicines today. Open access to the information would also encourage more scientists to work at the cutting edge of knowledge, to tackle diseases and disabilities more quickly.

And yet these social benefits would not impoverish the pharmaceutical companies. The companies would continue to be rewarded for their work, because instead of patenting their innovations, they would apply for remuneration rights, which would entitle them to payments from a central fund in proportion to the health benefits of innovative drugs – *regardless of who actually manufactured them.*

For in medicine what matters fundamentally is improving the health of individuals and populations. This may mean actually saving lives or simply improving the quality of life by reducing pain or avoiding disability. Our aim should be to ensure that the resources we dedicate to this are spent so as to maximize these improvements. We need, therefore, to track not only who uses particular medicines but also the estimated benefit to their health (at least on average). In order to tie payments to outcomes in this way, a standardized metric is needed to enable comparisons between treatments of different kinds. For example, how can one compare a treatment that fights a rare form of cancer and saves a hundred people under 30 with a different cancer treatment that saves two hundred people whose average age is 65?

One answer is by measuring what are known by the ungainly name of "quality adjusted life-years" (QALYs). These allow a treatment to be assessed according to the number of extra "life-years" it is presumed to have provided, and the quality of those years. So if a treatment saves the life of a 30-year-old who can be expected to live another 40 years, it has a value of 40 QALYs, whereas a treatment that saves a 65-year-old whose life-expectancy is only five more years has a value of 5 QALYs. In the case of treatments that prevent disabilities, of varying severity, the value in QALYs is weighted accordingly.[4]

Under the Open rules, the share of the remuneration fund paid to different rights-holders would be proportional to the health benefits, calculated as the number of people treated multiplied by the estimated benefit per patient in QALYs.[5] By relating remuneration rights directly to the benefits that various medicines bring, this system creates incentives for targeted and socially beneficial research. And at the same time, the lower prices of drugs gives patients dramatically expanded access to treatment.

An important consideration in these distributions would be the reuse of information that is itself covered by remuneration rights. Research is a cumulative process, and as the case of AZT vividly illustrates, innovations that yield new medical treatments usually build upon and incorporate previous work. It is important to think about how this is handled in the Open model, because otherwise the incentives might become heavily distorted, with resources going to the wrong people for the wrong things.

Consider the problem of derivative drugs that arises under the current patent system. Imagine that researchers at the company WorkedALot create a new drug for diabetes called Diax. They apply for and receive a patent. Then another company, DerivativesInc, produces a slightly cheaper variant called Diox, and this too is granted a patent. Clearly it would be unjust and would reduce the incentives to innovation if most of the rewards were to go to DerivativesInc at the expense of WorkedALot as it

---

[4]This system of valuing lives has been criticized as contrary to, for instance, religious teachings and the spirit of the UN Declaration of Human Rights, for which it is axiomatic that all lives are equal. These are genuine objections. The weightings in particular are obviously debatable, and would be subject to revision with advances in knowledge. But QALYs are not measures of the value of one individual's life; they are a statistical device, and they do at least provide a systematic way to assess the value of medical interventions. We inevitably have to make judgments about priorities (and already do), so it is better to have a measure of some kind than none.

[5]Refinements to this basic formula might be made to account for rare diseases where the number of patients may be small, for example by including a health prioritization multiplier in the formula.

is WorkedALot which did the pioneering and costly work. So under the present patent law, there are means for ensuring that such derivatives must license from the originator together with a dispute resolution system. This model would be adopted under the system of remuneration rights, with one major difference. With remuneration rights, lack of a licence would not prevent DerivativesInc from engaging in research or releasing its drug (though it would run the risk that later arbitration might award much of its future income to WorkedALot). Patents, by contrast, are usually interpreted as providing complete exclusion: without a licence, reusers are liable for damages and can do nothing – if they go ahead in the knowledge that they may be infringing the patent, they risk more severe damages for wilful infringement.

Opening all research, whether financed privately or by the state, may sound all very well for a single country, but it does raise the free-rider problem: if all the publicly funded research the US does is Open, won't others skimp on research and use that instead? What was to prevent South Africa's government relying entirely on American research into HIV and funding none of its own?[6] Well, even in an age increasingly obsessed with IP, there is a way to solve this problem and to refocus policy on innovation and outcomes rather than corporate protection.

The solution is international agreements under which countries commit themselves to minimum levels of medical research funding (as members of NATO, for instance, do currently in the case of defense spending). At its simplest, each country would agree to allocate, say, 0.5% of GDP, but it is more likely that the percentage (as well as the gross level) would differ between countries, with richer countries committing themselves to higher proportions. Countries might also agree to reciprocal recognition

---

[6]This problem is not limited to the Open model. It exists whatever approach one takes to paying for the production of information, including monopoly rights such as copyright or patents. For example, if a country does not recognize the patents or copyrights of its neighbour, it can benefit from the innovative and creative efforts next door without contributing to the cost: they use the products but pay a lower price that does not include a component for the rights-holder.

of remuneration rights, so that a remuneration right registered in one country would be rewarded also from the remuneration rights funds of other countries where a drug had saved lives or reduced suffering.

This is almost exactly the approach set out in the Medical Innovation Convention proposed by Jamie Love and others. As well as tackling the free-rider problem, such agreements could also allow more systematic international prioritizing of neglected areas. At the moment, more money is spent each year finding drugs to reduce signs of ageing than on fighting malaria. Yet more than half a million people die each year of malaria, whereas no-one dies of wrinkles. Because people in rich countries aren't affected by malaria but care a lot about ageing, drug research by big pharma is directed not towards deadly diseases but to the gratification of vanity. Through international agreements, it would be possible to focus research on these neglected diseases, for example by weighting research investments when calculating each country's spending commitment as a proportion of GDP.

So far, sadly, no such agreements on medical research have been negotiated, yet over the past decade there have been some large victories. When Jamie and others started work on access to medicines in the late 1990s, fewer than ten thousand people in the developing world were receiving effective HIV therapy. That figure is now above ten million, thanks to price reductions from the making available of essential drugs through voluntary, or sometimes compulsory, licensing of patents. Thousands, possibly millions of people are alive today because of work by Jamie and his colleagues on an area of information policy that most of the beneficiaries have never heard of.

Although the Medical Innovation Convention has not been adopted, it remains a blueprint for a different model of research funding. It and proposals like it are essential to the future of the information age, as ways to marry Openness with a resolution of the free-rider problem while continuing to benefit from market mechanisms and up-front funding. Such models rely upon international agreements, but so does the current solution to the free-rider problem, the granting of monopolies for "intellectual property", which were going to be extended yet again in the TPP (Trans-Pacific Partnership) – although this time Donald Trump's America did not sign.

11
Making an Open World

The principal advantages of the Open model are simple:

- Universal access to information
- Increasing innovation and creativity
- Maximizing positive use of the capacities of information technology
- Increasing competition
- Ending of global monopolies over various forms of information
- Reducing inequalities of opportunities and outcomes
- Increased global wealth

Together, these amount to an overwhelming case for the Open model, but there are still questions to answer about how it can be put into operation. These are not technological but political questions, often about our values and priorities, and this is the time for policy-makers and commentators to discuss them. In doing so we must keep in mind that the Open model need not be perfect. It only needs to better than our current Closed one, and sufficiently so to warrant change.

In the coming years, more and more people are likely to realize that monopolies in information have effects that are even more pernicious than monopolies in physical goods, particularly when it comes to inequality and innovation. "Intellectual property" will come under increasing criticism as its consequences become more visible in the virtual and real worlds.

However, any change must address the main concern people have about the alternative: how will we pay for the production of valuable information in an Open world where there are no copyrights and patents? The Open model provides a simple, comprehensive answer: replace current patents and copyrights with remuneration rights while maintaining existing funding sources that are compatible with Openness, such as government and philanthropic funding for research, and community-resourced projects like Wikipedia.

Today, information production is funded in a variety of different ways, as it will continue to be in future. As substitutes for copyrights and patents, remuneration rights will play a big part, but their scale and logistics need to be determined by research, discussion and planning. They will evolve along with technical and social concerns.

Huge amounts of information are created by all of us all the time – our blogs and photos, novels that never see the light of day, emails – but most of this is without economic cost or consequence, and will continue unabated. It is the funding of valuable information, whether Open or Closed, that matters here, and it has many sources:

- Business (journalism, film production, market research, advertising, fashion)
- Sponsorship (whether commercial or pro bono)
- Philanthropy (research, the arts, architecture, prizes, etc.)
- Crowdfunding
- State spending (universities, learned societies, charities)

All five of these forms of funding will continue under the Open model, and a good deal of the information they produce is already Open – or could be. For example, information created by publicly funded researchers is already largely Open – and the rest should become so.[1] But Openness is not principally about

---

[1] There are other examples of Open information production today. For instance, some innovations by commercial operations are Openly shared, with revenues being generated from complementary goods, related

state-funding or depending on volunteers. The more information we can produce without direct state control the better, because this minimizes politicization and bureaucratization and permits the greatest freedom for enterprise. The commercial business of making and marketing information must go on flourishing, and Openness can coexist with markets because we can use remuneration rights.

11.1   Remuneration rights in place of monopoly rights

The production of knowledge and information that is already Open today would be unaffected by the removal of "intellectual property". But there remains a huge amount of information which is paid for by businesses which rely upon patents or copyrights for their returns. What this book proposes is the wholesale replacement of intellectual property monopolies such as patents and copyright with remuneration rights. This would mean remuneration rights for software, statistics, design, news, maps, medicines, and a myriad of other kinds of information. Furthermore, we would replace each of the main intellectual property rights (patent and copyright) with a similar remuneration right: i.e. a patent-like remuneration right and a copyright-like remuneration right, each

---

services or consultancy. This is termed the "fries and ketchup" approach: give away the fries and sell the ketchup, or vice-versa. Pioneers in many fields, including some in which innovations are not recognized as "intellectual property", do likewise. A good deal of the innovation in, for instance, surgical techniques is not (and could not be) patented, but surgeons developing new methods often share them eagerly, knowing that this will bring them professional credit and more patients (nor should one overlook the simple desire to do good and improve human well-being). The new idea is shared freely, for indirect rewards, and this kind of innovation is much more common than we may at first suppose. Research by Eric von Hippel and his colleagues has shown that it extends far beyond medicine, from Michelin chefs to the chemical industry. Remarkably, von Hippel estimates that the majority of *all* production innovations are made by practitioners, a great many of whom do not seek or need exclusivity to justify or resource their efforts.

with the same qualification rules and term of operation as those of the respective existing monopoly rights.

Remuneration rights are entirely compatible with continuing other means of encouraging innovation – and can enhance them. For example, the existence of remuneration rights would provide new avenues for private philanthropic support. Philanthropists could for instance donate money directly to the remuneration rights pool, making the rewards larger for, say, photography or poetry. And private support for Open innovation through prizes or remuneration rights could be encouraged by tax breaks.

11.2   Are remuneration rights feasible?

In order to represent a viable alternative to the patent system, remuneration rights must be technically and politically feasible to implement. Technically, many of the aspects required under a remuneration rights system already exist; we already have means of measuring value, we already define ownership of innovations, as well as what happens when innovations are built upon by others. Each of these mechanisms could be reused for remuneration rights.

Furthermore, much of the political infrastructure required for a remuneration rights system is already in place, including coherent international (and often national) legislation and means of arbitration that could be co-opted, as well as similar governing bodies for related funds, and the means of securing sustainable funding.

11.3   Remuneration rights are technically feasible

For remuneration rights to be a viable funding mechanism, the technical aspects of the model must be practicable. Four issues stand out:

1. **Demarcation.** Which innovation belongs to which innovator?
2. **Reuse**. As research is cumulative, it is important that the remuneration rights model rewards innovators in a proportionate

manner, without disadvantaging those upon whose shoulders they stand.

3. **Distribution**. How should we allocate those funds to individual holders of remuneration rights?

4. **Evaluation**. How should we determine how much to spend on different *kinds* of information (medicines, music, software, etc.)?

Fortunately, there are precedents for all of these requirements.

11.3.1   *We already determine who owns innovations*

It is crucial to both the patent and the remuneration rights systems that innovations can be separated from one another. In order to give a right, whether a patent monopoly right or a remuneration right, to an individual, we have to be able to attribute innovation correctly. This process is vital in the patent system, and could be directly reused in a remuneration rights system.

11.3.2   *We already share rights between multiple innovators*

Remuneration rights would be granted on the condition of completely Open access to all information relating to the innovation. It is therefore important that remuneration can be shared fairly between one generation of innovators and the next.

This kind of sharing already happens. Because innovation and creativity are cumulative, reuse is often frequent. Under today's monopoly rights system, follow-on innovators are required to pay royalties to the earlier innovator. Under the remuneration rights model, in a similar fashion to royalties in the patent or copyright system, follow-on innovators would be liable to pay a proportion of their own remuneration rights payments to those whose work they built upon. These proportions might be standardized for simple cases or, for more complex cases, the two parties could negotiate, with ultimate recourse to the courts if no mutually acceptable solution were found. In other words, if an innovation built upon a previous innovation holding a remuneration right, then a proportion of the right granted to the secondary innovation would be set aside for the primary innovator.

The major difference with the present system would be that earlier innovators would not have an absolute right to prohibit reuse as they do today. Rather, they would have the right only to "equitable remuneration". This change would favour the succeeding innovator but still ensure that the earlier was fairly compensated.

11.3.3   We can distribute funds between holders of remunerations rights

Paying creators from a remuneration rights fund is fairly straightforward, and the case studies above have already demonstrated how it would be done in music and medicines. The distribution is done by comparing similar things, which allows a common yardstick, whether it be the number of plays of different songs or the usage and health benefits of medicines. Whilst the fund for each kind of information would need its own specific mechanism (software would have a different criteria from music, for instance), the key principles are clear: holders of remuneration rights would be paid in accordance with usage and the value created by their innovations; and distributions would be made by transparent, pre-defined algorithms overseen by an independent assessors (so as to eliminate the risk of political meddling in the process).

11.3.4   We can evaluate how much should be spent on each kind of information

The proportionate allocation *among* the remunerations funds is a greater challenge, because it requires comparison of values that are incommensurate: how much do we value a new single by Beyoncé compared to a new treatment for breast cancer?

We can begin by looking at how we make such judgements when it comes to dissimilar physical things such as a football and a cake. The most common mechanism is the market. The interplay of buyers and sellers determines prices (and therefore a form of relative value), as well as how many footballs and how many cakes are made. But while traditional market pricing works well when we want to compare physical things, because there is a limited supply of each, this mechanism breaks down in the case of information. Because as we have seen, there is no limit to the supply of digital information unless we deliberately restrict it. As economists might say, the cake gets larger as required. Everyone can have a slice and no one need go hungry. In a market system for information *without* monopoly rights, this limitless supply would result in prices tending to zero – giving us no indication of relative values of different types of information.

But what about monopoly rights: don't they give us market prices for information goods? The issue here is that market prices are useful not because they exist but because they allocate spending and production in line with actual value and costs. However, this is so only under certain conditions that are largely absent in the case of information goods covered by monopoly rights.

First, information goods (in common with public goods such as national defence) have significant one-off costs but trivial costs for each additional user. For example, once we have spent the money to maintain an army, protecting an extra hundred citizens comes at no cost. Similarly, once an app is created, the cost of an additional copy is zero. This creates challenges for a market pricing system as the large fixed cost means that prices fall as use rises (although the value is not falling).

Secondly, at a fundamental level, true market pricing of information goods is impossible. Only when the state creates artificial monopolies can prices be attached to information – but the very act of doing so undermines the mechanism of the market by allowing the producer, rather than consumer demand, to set the price. Although there may be other films to watch, none of them is exactly equivalent to *Harry Potter* or *The Italian Job*, so there is no true competition. The mechanism is distorted and as a result resources are not optimally allocated.

Nor is the regime of "intellectual property" the pure free market mechanism that it is often taken for: it is already politicized (which is why it attracts so much lobbying). For example, the length and nature of the monopoly rights that society grants to particular kinds of information depend upon political decisions. Should patents for new life-saving drugs last for 10 years, or 20, or 50? (Think of the profound implications of such a ruling.) Governments have to decide, too, how much to spend on scientific research, where there are even more factors that make evaluation highly speculative. Basic biological research, for instance, may or may not lead to medical breakthroughs, but with a time-lag of decades.

So the challenges remuneration rights face in assessing value and allocating funds between different types of information are *already* present in the system we have today – albeit more hidden from our view. The issues that remuneration rights raise are not new, just more visible.

Economists do have techniques to assess value, and so to compare – albeit imperfectly – musical apples with medicinal oranges. Sampling techniques, for instance, give a good idea of the overall usage of information goods, be they apps, weather forecasts or algorithms. This tells us only the levels of use, not the value that the users put upon it, but there are also techniques such as willingness-to-pay surveys. So-called "hedonic pricing" uses things that have prices to evaluate things that haven't. For example, what is the value of a beautiful view? There is no explicit market in beautiful views, but we do measure the prices people pay for houses. So if in addition we know which houses have beautiful views – along with other factors such as size and location – we can begin to tease out the implied value of a beautiful view. We can even say something about the value of life. What will people pay to avoid a small increase in the risk of death? For example, what bonuses do we have to pay people to take on dangerous jobs such as cleaning the windows of skyscrapers?

Furthermore, we can initiate the remuneration rights regime using the knowledge we have to spending levels today: how much we spend on music versus movies versus software versus medicines, etc. Whist imperfect, these existing expenditures do provide a useful guide, and basing remuneration rights on them would also provide a welcome continuity.

In conclusion, whilst there can be no final answer to the questions of allocation, it is possible to suggest a general approach to them in an Open world. In each creative field the current level of investment offers a starting line for the new model, and we can at least put nominal values upon our information goods in the Open world, using existing tools and the large amount of data available about the levels and forms of usage of digital goods.

Ultimately the sums to be channelled to different forms of information through remuneration rights are necessarily matters for public debate. How much should we allocate to the new information in the many different realms? But the challenge of setting levels of expenditure is neither new nor specific to information. Societies face the same problem when deciding how much to spend on parks, schools and fighter planes.

This is why we need to start discussions now about how large the remuneration funds should be, how they should be collected, and how they should be distributed. We also need to explain to the public what remuneration rights are, and the benefits that Open access will bring both to society as a whole in the form of cheaper medicines, faster advances in research, and so on; and to each of us personally in the form of cheaper, non-proprietary goods of all kinds; a much freer internet with universal access to news sites, music, films, books and much more.

11.4   Remuneration rights are politically feasible

As well as functioning technically, remuneration rights must be able to operate politically. This involves

1. Adequate and sustainable financing of the funds
2. A robust governance structure and legal status for the funds
3. A successful transition from where we are today to the new Open model

   These requirements too can be met.

11.4.1   *Sustainable funding can be ensured, nationally and globally*

Starting at the national scale, governments must provide a predictable and reliable level of resourcing. There are several ways this can be achieved and these will necessarily vary with the existing circumstances and practices of the country in question.

The funds should be legally independent, with transparent governance. This "ring-fences" the money, keeping it separate from the general government budget. The most important feature of the governing body would be impartiality. One way of guaranteeing this would be to make it independent of electoral politics and political factions.

On the global scale, international agreements must be reached, establishing a system to set equitable contributions to the fund and binding all countries to contribute. This would ensure the fund would have a fixed disbursable pool each year. Only by establishing such binding agreements for specific types of remuneration rights can free-riding be deterred.

Happily, we already have such mechanisms in place. For example, the existing intellectual property regime demonstrates the effectiveness of international agreements to prevent free-riding. And we already have many examples of international initiatives where governments pool funds, for example in research funding or space exploration.

11.4.2   *Remuneration rights are compatible with national and international laws*

Most countries are signatories to treaties such as the World Trade Organization's TRIPS, which require provision and recognition of patents and copyrights. The remuneration rights model is compatible with these legal frameworks.

Whilst TRIPS is global and binding, it also has built-in flexibilities. Legal provisions are in place to allow exceptions to exclusive rights in order to widen access and to bypass monopoly rights under specific circumstances. For example, "compulsory licensing" is permitted in certain circumstances to ensure that patent owners cannot block the use of their innovations – though they must be suitably compensated.

Temporary difficulties in the process of completely replacing monopoly rights need not prevent progress. For example, holders of monopoly rights could voluntarily license their rights into a remuneration rights fund – just as copyright holders do with collecting societies or Spotify today. Alternatively, remuneration rights could be granted in parallel with monopoly rights, with innovators having to choose one or the other. Remuneration rights could then be made more attractive than monopoly rights in a variety of ways to ensure take-up: for example, by providing high levels of funding for remuneration rights; or by giving preference in all state spending to remuneration rights (state spending on patented drugs, for instance, could be limited); or finally by directly taxing income from patents, so reducing their attractiveness (but without breaching TRIPS and other international agreements). Although this opt-in approach is less attractive than a full transition, it might be useful where abolition of monopoly rights is not feasible in the short term, for political or legal reasons.

11.4.3   *We can make a successful transition to an Open model with remuneration rights*

Obviously the quickest method of introducing the Open model would be a global "big bang", with all existing monopoly rights being abolished overnight, replaced by remuneration rights. Equally obviously, this is unlikely to happen because of the scale and complexity of such a change. Instead, we should encourage incremental adoption, both by region and by industry. Individual nations or groups of nations can adopt the Open approach while other countries retain monopoly rights. And it is quite feasible for one country or group of countries to adopt the Open model initially just in one or a few industries: for example, introducing remuneration rights for, say, music but keeping monopoly rights for everything else. This ability to pilot the Open model, and to run it in parallel with the existing monopoly rights system, is a huge advantage. It both allows for testing of the new approach and for the gradual adoption essential to the success of such a collective effort.

Another key requirement for any change is that major interest groups will support it – or, at least not actively oppose it. Given their political power, a key group will be existing holders of monopoly rights such as pharmaceutical companies, record labels or publishers. Such interests are always wary of any change to the status quo, but there are reasons to be confident that remuneration rights can offer advantages to them as well as society. First, from the point of view of a monopoly rightsholder, a remuneration right looks quite similar: it will be issued by a similar body, last for a similar period and yield a similar or greater income. This, of course, is the crux, and it is essential that the amount of money guaranteed through remuneration rights funds compares favourably with current income from sales. For instance, if the US set up a remuneration rights fund for medicines, it should be *at least* as richly endowed as the total pot of money spent on patent medicines today. The companies need not lose out, even though the public can be greatly benefited by increased access and lower costs.

That is the beauty of the win-win Open Revolution, and together, these features of remuneration rights greatly increase the chances that this change is politically as well as technically feasible.

12
Help us Make it Happen

Concerted action is required if we are to create an Open world. Some people assume that because digital technology makes it so easy to share freely, nothing can hold back the free flow of information. But optimism about the inevitability of Openness is naive about the interplay of technology and power. For power, whether that of special interest groups or us collectively through the state, does much to shape and control the impact of technology, especially in this area of information. Even if digital technology did create a world in which information could not be prevented from flowing freely, there would still be the question of who pays to create it in the first place. In the absence of a new financial structure such as proposed here, free information might well *hinder* innovation and creativity, by robbing innovators of their income, so impoverishing us all.

Another mistaken assumption is that if we leave the free market to work, the Open business model will prevail on its own. Alas, whilst Open business models in a Closed world (or community efforts such as Wikipedia) are impressive, they can only pay for a very small portion of the information we want and need. Many important information goods – such as new medicines or new films – have no obvious Open business model in a Closed world without remuneration rights.

Action is therefore needed and it should take three complementary forms: informing and involving the public; lobbying for policy change; and building-it-ourselves. Ultimately, we need to change policy at a national and then international level. At the same time, individually or in groups we can take action now to create Open materials, whether software, databases or content with the intention of both delivering immediate value and acting as exemplars of the potential of Openness.

A fundamental shift in the public conception of information, recognizing the benefits of sharing rather than hoarding, needs to be matched by political pressure to fund Open information; to establish disbursement processes, including the creation of remuneration rights as legal entitlements and the mechanisms for licensing and dispute settlement; and to make international agreements instituting Open policies.

But before policy changes can be brought about, we need a broad-based Open movement with a common language and goals: both are currently lacking. This movement requires a vanguard of individuals and organizations engaged in advocacy. For inspiration, they can look to other efforts to secure major governmental changes in the public interest, for example the environmental movement.

Concerns with problems such as pollution go back at least to Roman times, but in societies that were primarily agricultural, the human impact on the wider environment was barely noticeable. With the coming of the industrial age environmental matters started to receive greater attention, boosted by the Romantic concern with the sublime, the beauty of nature, and the ugliness of manufacturing cities. As industry and urbanization grew, the pollution of water and air became starkly visible, but regulation was limited for a variety of reasons including poor understanding of the science and political systems which favored owners over workers. Even in the 20th century, environmental progress was hampered not only by two world wars, but by a political failure to think long term.

After 1945 the momentum of environmental concern finally picked up, fueled by better science, maturing representational democracy, and states taking a more active interest in general social welfare. Increasing wealth raised the relative value of environmental goods such as parks, clean air and long-term health, as well as the leisure to enjoy them and the willingness to vote for them. In the early 1960s these crystallized into the modern environmental movement. The symbol and catalyst of this was the publication in 1962 of Rachel Carson's *Silent Spring*, which brought home to middle America the alarming impact of man-made chemicals on the environment and human health.

It would take decades for the environmental movement to mature, but by the early 1990s, several environmental organizations had become significant political forces, with membership in the hundreds of thousands and substantial funding. They had independent researchers, sophisticated media strategies, grass-root campaigns, and lobbyists in major political centres. Though still vastly outgunned financially, Big Environment had arrived to take on Big Business. Public awareness had also grown, and terms such as "sustainable" and "green" had entered the popular vocabulary. By the turn of the century, calling a car or a house "green" was usually no longer a description of its colour.

Meanwhile, though, the scale of the damage had greatly increased, and people had become increasingly alert to ever larger challenges, with climate change the greatest of them all. Yet oil producers, whether ExxonMobil or Saudi Arabia, remain among the most powerful lobbies on the planet, and continue to fight every inch of the way to prevent action on climate change that will reduce their profitability.

The analogies with the information environment are striking. Concerns about the control of information also go back hundreds of years – think of the Church's refusal to allow the Bible to be translated, lest people read it for themselves. Again, though, it was with industrialization that information goods, ranging from manufacturing equipment to newspapers, became economically and socially crucial. Here too, though, political discourse was dominated by special interests, especially those of the producers and controllers of information. Mechanisms for representing broader groups such as consumers were poor and although information was becoming increasingly commercially valuable, few people understood the conceptual significance of this. Closed monopoly rights grew in term and scope.

In the second half of the 20th century, however, two new factors came into play. The invention of digital technologies led to the proliferation of information and to costless copying; and the growth of Big Science, with research heavily funded by government, massively expanded the production of information in the public sphere. Yet still, with information increasingly dominant in the economy and society, there was little political understanding. Elements of Open sharing were widespread both in the academy and in the nascent information technology industries, especially software, but this was rarely driven by political conviction.

Finally, in the 1980s and 1990s, the Open information movement appeared in embryonic form. If one were to look for a totemic moment similar to the publication of Carson's *Silent Spring*, it would probably be Richard Stallman's work at the Free Software Foundation. Whilst Carson exposed the harm done by pesticides, Stallman revealed the increasing threat posed by the way that more and more information was becoming proprietary. At first on a tiny scale, a community of coders and scientists, connected by personal computers and the rudimentary internet, gathered around a radical ideal, as they gradually worked out the ramifications of the difference between information and physical things.

Information politics began with such groups as the Free Software Foundation (founded 1986), the Electronic Frontier Foundation (1990), and the Foundation for a Free Information Infrastructure (1999). There were some major early triumphs, such as the rejection of software patents in Europe in 2005. Groups formed and re-formed, but there was little shared language or vision. The excitement of the internet economy spurred consideration of monopoly rights and the potential of Openness in an information age – most notably and most popularly by American law professors such as Lawrence Lessig and James Boyle. At this stage, the corporate opposition to Openness was powerful but still cumbersome and unsophisticated.

Public science was also increasingly aware of the need for Openness as research began to become entangled with commercial interests and monopoly rights – as evidenced in the case of the Human Genome Project, which began in 1990 with funding from both government and philanthropists. Eleven years later, when the project released the first ever complete sequence of the human genetic code, it was completely Open. Now it is the basis of a global industry worth more than $20 billion, and its Openness is far from unusual: today more than half of all basic medical research is released Openly. In February 2018, an even more ambitious project began to be planned by American and Chinese researchers. The BioGenome Project (or Whole Earth Genome) aims to sequence *all* species on earth, from amoeba to blue whales. The cost would be in billions of dollars, and the sources of funding are so far unclear, but it is obvious to politicians as well as scientists that such an extraordinary databank must not be held exclusively, but shared as widely as possible with everyone who might make use of it.

More than half the planet is now online and in regular direct contact with the world of digital information. And yet, no common goal or language of Openness exists. If you were to stop people in the street and ask "Do you want an Open world?" or even "Do you want an Open information society", they would have no idea what you meant, just as if you had stopped them in 1975 and asked "Do you want a green society?" The Open information movement is still incoherent, unstructured and without an acknowledged global spokesgroup, a shared purpose or a platform or an agreed approach. "Information politics" is a term barely understood. The published manifestos of political parties scarcely mention it, and when they do it is usually only to reiterate dogmas about innovation and intellectual property rights.

Of course, individuals, groups and businesses are already creating Open information within the paradigm of monopoly rights. There are volunteer efforts motivated by a combination of personal interest, public-spiritedness and a desire to develop and demonstrate knowledge and skills (examples include Wikipedia

and many small Open software projects).[1] There are also businesses that are providing Open information for free while selling
complementary goods. Philanthropic support is a source of increasing funding in many areas, though uncoordinated and not
yet systematically promoting Openness.

Yet even as the public wakes up to the excessive power of
those like Facebook and Google, many still fail to understand the
constituents of these monopolies. They do not understand that
it is the *rules* we have made that create and sustain them, that
it is "intellectual property" monopolies that are allowing these
extraordinary concentrations of power and wealth. We must go
on spreading awareness and challenging mistaken conceptions
of the workings of the digital information economy. We need a
world where every policy-maker, every expert, every educated
citizen understands that bits are different from bread – and what
that implies.

We need to spell out the dangers of a Closed world built on
proprietary information. We need research bodies dedicated to
understanding our information economy and society, as well as
think-tanks to track progress and develop policies. We need mass
membership organizations to campaign for an Open world, in the
way that Green organizations campaign on behalf of the physical environment, because membership provides the resources to
engage long term and a clear constituency supporting change.
Finally, we need policy-makers to see the opportunity, and neces-
sity, of an Open world. For, ultimately, it is only by our collective,
political action that we can effect the large-scale reforms we need.

*To find out more about what we can do to make an Open world
and how you can get involved, visit:*

https://openrevolution.net/make-it-happen

---

[1] Wikipedia is a hugely impressive voluntary effort, but it was kick-started
commercially and it too has benefited from state spending: its content is
largely collected from information already published elsewhere, much of
it produced in state-supported academia (or in commercial contexts such
as journalism). One way or another, a good deal of the Open material
available to us today has been supported by governments and business.

13
Coda: The Original Copyfight

Fifteen hundred years ago, in 6th-century Ireland, a dispute over the copying of a book led to a pitched battle. At its heart was a priest named Colmcille, better known to us as St Columba. After ordination at 25, he began travelling around Ireland and founded three dozen monasteries in 15 years. With shortages of books restricting religious scholarship, he copied manuscripts whenever he could, and encouraged his monks to do the same, to spread the Church's teachings.

The Vulgate was St Jerome's great 5th-century translation of the Bible into Latin. The first copy to reach Ireland was brought from Rome by Finnian of Molville. Although so protective of his book that he would not allow others access to it, Finnian had been one of Colmcille's teachers and he made an exception for his pupil, allowing him to read the translation so long as he did not copy it. Colmcille acquiesced, but ignored the restriction, no doubt feeling that this knowledge was too precious to be locked away.

He proceeded to copy the book at night as fast as he could. Discovered one night, Finnian demanded that he hand over the copy. Colmcille refused, believing that the holy text could not be owned by anyone, and that his first duty was to God and the Church, not to Finnian.

When Diarmid, the High King of Ireland, was asked to adjudicate, Colmcille made his case by saying that it was the duty of the Church to spread its knowledge by copying. He had not diminished Finnian's book by doing so, and in any case, that too was a copy, since it was not St Jerome's original manuscript. Books were different in kind from material goods.

King Diarmaid, however, ruled against Colmcille: wise men had always described the copy of a book as a child-book, he said, which implied that the owner of the parent-book also owns the child-book. "To every cow its calf, to every book its child-book. The child-book belongs to Finnian."

Colmcille is said to have cursed the King and stormed back to his monastery, and soon afterwards events took a bloody turn, when Colmcille granted sanctuary to a hostage who had been taken by King Diarmaid. The King violated the sanctuary of Colmcille's monastery by having the hostage hunted down and killed. A battle ensued in which, legend has it, three thousand of Diarmaid's men and only one of Colmcille's men were lost.

But Colmcille's victory was short-lived. He was excommunicated by a synod of fellow priests, and then exiled. So in 563, two years after the "battle of the book", Colmcille set sail with 12 followers for Iona. He went on to found a new monastery and to play a major role in bringing Christianity to the Picts in Scotland.

Colmcille's understanding that the text of the Vulgate could not be property in the usual sense was, of course, based upon its doctrinal importance rather than a modern sense of information as the basis of the economy, scientific progress and much else. Yet his distinction between ownership of physical possessions and the intrinsic replicability of information is even more important a millennium-and-a-half later. His final, passionate plea before the King had perfectly articulated the essential logic of Open information:

> My friend's claim seeks to apply a worn out law to a new reality. Books are different from other chattels and the law should recognise this. Learned men like us, who have received a new heritage of knowledge through books, have an obligation to spread that knowledge, by copying and distributing those books far and wide. I haven't used up Finnian's book by copying it. He still has the original and

that original is none the worse for my having copied it. Nor has it decreased in value because I made a transcript of it. The knowledge in books should be available to anybody who wants to read them and has the skills or is worthy to do so; and it is wrong to hide such knowledge away or to attempt to extinguish the divine things that books contain. It is wrong to attempt to prevent me or anyone else from copying it or reading it or making multiple copies to disperse throughout the land.

14
Acknowledgements

My deep gratitude to Sylvie for everything she has contributed to making this book happen, and my fulsome thanks to my editor Jim McCue who helped wrangle the book into a coherent form.

As this book attests, our culture and our ideas are built on the shoulders of others. Innumerable people have shared their thoughts and time with me over the years and this book rests on their contributions. There are many more than I could adequately acknowledge and here I can only begin to thank some of those who directly contributed to the making of this book: Lionel Bently, Tanvir Chahal, Shelly Chen, Michiel De Jong, Cory Doctorow, Lottie Fenby, Ninon Godefroy, Jonathan Gray, Robert Hart, Tim Hubbard, Laura James, Adam Kariv, Liam Kavanagh, Martin Kretschmer, Geoff Mulgan, Lieke Ploger, Cressida Pollock, Karen Pollock, Gordon Pollock, Allison Randal, Andrew Rens, Esteban Ruseler, Philipp Schmidt, Tom Steinberg, Audrey Tang, Fiona Thompson, Paul Walsh, the Open Knowledge community, the Shuttleworth Fellows and many, many more.

Made in the USA
Las Vegas, NV
20 May 2021

23382793R00075